Information Technology and Development

By

Dr. M. Lakshmi Narasaiah
M.A., Ph.D.
Professor & Head
Department of Economics
Sri Krishnadevaraya University Post-graduate Centre
Kurnool–518 002
Andhra Pradesh (India)

DISCOVERY PUBLISHING HOUSE
NEW DELHI

First Published–2004

Reprint : 2012

ISBN: 81-7141-760-4

Published by:

DISCOVERY PUBLISHING HOUSE

4831/24, Prahlad Street, Ansari Road, Darya Ganj

New Delhi–110 002 (India)

Phone: 23279245, • Fax: 91-11-23253475

e-mail: dphtemp@indiatimes.com

Printed at:

Dynamic Printers

Preface

Galloping advances in information technology promise to give us instant access to all world's knowledge. But how will human memory fare against the rise of the super-machine? If the architects of technology's next great leap forward are to be believed, all knowledge may soon be shrunk to vanishing point. Nanotechnology, or computing carried out at the scale of atoms, is their byword for the future. With its awesome potential, scientists have recently argued around 11 million 400-page volumes could be stored and primed for instant viewing on device the size of a human palm.

The ink may still be fresh on these blueprints, but the elixir of portable omniscience no longer seems so far away. Seemingly cast-iron laws of ever increasing computer power, along with the rise of powerful new technologies, appear to point to horizon where all that can be known and remembered can be transferred to machines with which human beings then interact at will. And it is a future that for some is already spelling big trouble for the brain.

Surveys Point to Yawing Gaps in General Knowledge

Computers not only distract us from contemplation of deeper valleys; they discourage from contemplation itself. As surveys repeatedly show, knowledge of history, literature, geography and even current affairs seem to be on a sleep decline: 60 per cent of adult Americans cannot recall the name of the president who ordered the dropping of the first atomic bomb, just as 77 per cent of young Britons are perplexed by words Magna Carta. The day of the nano-shrunk library could soon some, but will any of its users be able to remember a single line of poetry?

The connection between these yawning gaps in general knowledge and the information technology is by no means established, but a host of thinkers in different fields are sure that issues is one that will shortly become all too pertinent. At the same time as it helps us and extends our physical capabilities, it diminishes our individual faculties. This is a vital question, one which has been around for a long time.

Good or evil, writing has nevertheless formed one of the main tools in the evolution of human memory. Indeed it is civilization's unrelenting hunger for placing memory in external stores–cave paintings, then manuscripts, libraries, printed works and finally computers – that has supported the entire march of the species. Each of these new technologies has helped humans "off-load" their memories. Pre-literate societies, for instance, depended on oral tradition for their expertise–a practice undermined by the flaws of overworked brains, though fertile ground for epic poetry. Through the written word, memories were freed from the head: knowledge could be stored for retrieval in books, and then redrafted into the sort of novel and complex codes on which modern society is founded.

Becoming Good Memory Managers

The benefits of storing memory outside the brain are unquestionable, but the invention of printing over 500 years ago followed by the post-war onset of computing have added a new note to the process: that of thundering acceleration. One simple equation has come to embody this. It stipulates that computing power–defined in terms of capacity and speed per unit cost—doubles every two years. The trend has held for the last 40 years. Should it continue as expected to around 2020, a personal computer by that year will have exactly the same processing power as a single human brain. Add the promised marvels of nanotechnology, optical and quantum computing, and machines might reach utterly daunting proportions. One penny's worth of computing circa 2099 will have a billion times greater computing capacity than all humans on Earth.

For many cognitive scientists, relations between mind and machine are already undergoing drastic reconfiguration. "Distributed intelligence" is the new maxim, encapsulating all systems in which individuals and computers mesh to carry out a collective task, whether it be landing an aircraft or tracking share prices. The Internet is so far the crowning glory—a system that in principle might combine individual users into a potent group min.

All of this may sound abstract, but the effects on memory are being felt now. Facts and figures no longer take pride of place in school curricula. Within the past two years, South Korea, Singapore and Hong Kong–havens of rote learning–have debated plans to axe huge swathes of standard classroom study. Experts in education stress that students must learn to be adaptable, skilled in manipulating symbols, able to respond to new situations; in short, ready to deal with the new economy, a realm where the computer is king.

Dr. M. Lakshmi Narasaiah

Contents

1

The Electronic Gap

In the United States, business journals, market gurus, economics professors and highly paid consultants talk incessantly about the coming global boom, the transformation of the workplace, the technology revolution and the knowledge explosion. It is implied that the world is slowly becoming a reproduction of Silicon Valley. It is asserted that this is the future. But instead of swallowing this hype, perhaps we should pull back and look at the globe as a whole.

The pessimistic view would be to point out what is currently occurring in Kosovo, West Africa, Rwanda, Chechnya, Kashmir and elsewhere. We might also follow Robert Kaplan in the trips he describes in *To the Ends of Earth,* only to discover that much of humanity is headed for disaster and self-destruction. I do not wish to be as negative as that. However, I would like to offer a caution to those who portray globalization in an uncritical and overly enthusiastic manner.

One in three Americans are regular, daily internet users. Even within American society, the computer and e-mail have widened the gap between educated people (chiefly whites and Asians) and the less educated (chiefly black Americans). This gap will be felt in every aspect of life, whether it is in opportunities, potential, education or job-hunting. The United States will be divided into two groups, one which is computer-literate and the other which is not.

This phenomenon has been replicated at the international level. The most important fact is that we are in the midst of a technology revolution that sees less likely to close the gap between rich and poor countries than to wider the gap even further.

The technology revolution and the communications revolution still bypass billions of human beings. The internet may have more influence than any single medium upon global educational and cultural developments in the coming century. Yet only 2.4 per cent of the world's population is one the internet, or one person out of 40. In Southeast Asia, only one person in 200 is linked to the internet. In the Arab states, only one person in 500 has internet access, while in Africa only one person in a 1,000 is an internet user. This situation will not change as long as those lands lack electricity, telephone wires and infrastructure. They cannot afford either computers or the expensive software they require. If knowledge indeed equals power, the developing world may have less real power nowadays than it did 30 years ago, before the internet was developed.

If we want to work toward a knowledge-based society in the coming century, over at least the next 10 years we need to make a concerted effort to bring poorer societies into the system of electronic communications. This effort will need to be coordinated by the World Bank, the UN Development Programme, UNESCO, the NGO community and the global business community.

The alternative is to perpetuate a world is fundamentally undemocratic and structurally unsound. If we do nothing, if we let the knowledge explosion intensify in technology in rich societies while poorer societies fall further behind, the growing gap between haves and have-nots will lead to widespread discontent and threaten any prospect of global harmony and international understanding. This is the most significant challenge we face. We have no time to waste in responding.

2

The Dot.Bomb Syndrome

Everyone agrees that privacy is good for business. But is there a market for companies offering to protect it? While several firms have recently suffered class-action lawsuits and public outrage over apparent intrusions into customers' personal data, American Express is one of a number of businesses that may be blazing the right path from privacy to profits.

Last year, the credit card issuer rolled out "private payments", enabling its customers to go online and obtain "disposable" credit card numbers that could be used for only one online purchase if the number was ever stolen after being used for purchase, it would be worthless.

A key lesson is that protecting privacy, in this case the credit card number, became an "enhanced value" to an existing service. Moreover, American Express understood that the credit card number was the "tip of the privacy iceberg" – it plans to roll out a suite of related services, like anonymous internet browsing, in 2001. Rival companies like Visa/ Mastercard are following the same trail.

Another company to watch is PrivaSys, of San Francisco, which has patented technology to enable plastic credit cards to generate disposal numbers for each purchase. PrivaSys does this by equipping the card with a calculator-styled key pad and LCD screen, a very thin battery, and a special magnetic stripe. The card holder punches a 4-digit PIN into the credit card itself and—Voila!—the card generates the disposable number.

But securing payments is just one part of an emerging privacy protection market. Other firms are offering anonymous or "pseudonymous" browsing tools, packages to control cookies—the strings of code that are planted on the user's computer by websites—and services to block hacker intrusion. Only time will tell which companies will control niche markets or achieve critical mass.

What's clear is that the nascent e-commerce industry badly miscalculated the importance of privacy to their business model. For without privacy, there was no consumer confidence, and without such confidence, e-commerce had no chance of fulfilling the short-term, high expectations that initially sent their stock values soaring. The current shake-out of the "dot.bomb" industry shows how costly this miscalculation over privacy was.

An Added but Essential Value

But if you thought privacy was a big issue for e-commerce, consider the debate on the wireless industry. Referred to as Mobile or M-commerce, this industry appears to have unlimited potential to deliver information and location-based services, advertisements, and discount coupons to cellular phones and hand-held wireless devices M-commerce risks hitting the major "privacy buttons": constant surveillance via location-based tracking; unsolicited ads clogging communications devices; detailed profiles of individuals' interest and movements; and insecure payment mechanisms.

Nonetheless, the industry is showing sings that it has learned form the mistakes of e-commerce, A few months ago a workshop of the U.S Federal Trade Commission, leading wireless industry groups unveiled guidelines requiring companies to obtain a consumer's consent before collecting or using personal data. This wasn't altruism: the industry understands that consumers will not tolerate commercial services to their cell phones unless they explicitly consent.

So far, privacy's brief tenure in the commercial market place indicates that some people are willing to pay extra to protect their privacy. But the one important lesson is that consumers are unlikely to take more than a few extra steps to secure their data. If they perceive that a given medium, like the internet is not privacy-friendly, they either change their habitual uses or actually refrain from them altogether. The likelihood is that privacy will became an "added value", creating a market for those who can integrate privacy into payment mechanisms without excessively burdening the consumer.

3

E-Learning—Designing Tomorrow's Education

The impact of information and communication technology is bringing gradual but often radical change – to working, to learning and more generally to our way of life. Europe has to make the best use of the new technologies. This is particularly crucial in the education and training. Yet schools are often still inadequately equipped with computers and internet connections. Technology – related skills among the staff concerned, in particular teachers and trainers, need to be improved. There are significant divergences between countries and regions of this globe. In this situation future is at risk in the increasingly knowledge-driven global economy.

E-Learning

Our definition of 'e learning' is the use of information and communication technology, including the internet, to learn and teach.

It also designates an initiative to:

- encourage the development and the acquisition of digital literacy:
- improve everyone's capacity to use new technologies for learning and working;
- adapt our education and training systems to meet the challenges of the information society.

The general objective—to make Europe a major player in the 'new economy'—was agreed in March 2000 by the EU Heads of State of Government in Lisbon, elearning falls under the European Commission's eEurope action plan' and 'strategy' for employment in the information society. These aim to bring Europe into the digital age.

The objectives

- To generalize and improve access to hardware, software and to information and communication networks.
- To provide and simplify access to quality training for all of us.
- To develop cooperation between teachers, trainers and managers involved in establishing an educational area.
- To gather and share information on the best practices based on the use of information and communication technology fôr learning.
- To promote innovation, know-how and expertise.

With whom

The e-learning initiative is open to cooperation with all interested parties and specialists in education and training, including;

- experts in the use of technologies in the areas of education and training;
- teachers, trainers and project leaders who develop innovative teaching practices in schools, universities and in all other educational establishments;
- senior administrative staff in the education and training sectors;
- pupil trainees, students, all those who want to learn with modern methods.
- information and communication technology companies, multimedia publishers, broadcasters;
- National, regional and local politicians interested in developing e learning initiatives.

4

Wiring Up the Ivory Towers

Prestigious universities are forging alliances to conquer a share of the e-learning market and stand up to virtual competitions just like airline companies, universities around the world are forming partnerships and consortia in response to the pressures of globalization. The World Education Market held in Vancouver was a timely sign; the fair, expressly organized to foster relations between universities, training providers, software companies and representatives from nations with large education needs attracted participants from over 60 countries.

This race to "partner up" is fuelled by a number of factors. In most industrialized countries, government funding for higher education has decreased, forcing institutions to look for new markets either to subsidize campus programmes or just to remain viable. There is a growing need for lifelong learning as "jobs for life" vanish and the information society drastically reduces the shelf-life of almost any educational qualification. Technological developments, increasingly necessary for learners in all fields to master, offer ever more innovative tools for supporting e-learning.

For business, online learning is "the" new market opportunity with the need for re-training and professional updating predicted to crease an $11.5 billion industry by 2003. Business is better able to develop and maintain the technological infrastructure necessary to run large online systems and everyone, including the universities, recognizes that is takes robust telecommunications technology to deliver education and training on the scale demanded.

A host of companies has sprung up to help universities shape and package courses for online presentation, while network providers are jockeying for position to deliver online education.

The United Sates is the undisputed leader in the field, prompting governments in the U.K., Canada and Australia to commission being eroded by U.S, ventures turned global,. Canada and the U.K are in the early stages of setting up their own virtual universities. But what has become clear is that the conservative and labyrinthine decision-making processes which characterize most university procedures are being jolted by a race to get a share of lifelong learning market.

So far, the most common approach for universities to break into the e-learning universities has been to develop courses specifically for a corporate partner or to form alliances among themselves. Universities 21, a company incorporated in the U.K, is a network of 18 leading universities in ten countries.

Very often, prestigious universities have stayed clear of going fully online seeing a danger to their brand name. Many are limiting their offerings to continuing education programmes and/or non-degree course and more often than not they are aiming at the corporate market. One company UNext.com has partnered with first-class institutions such as the University of Columbia (U.S.) and the London School of Economics to create online courses marketed under the name Cardean University. Their target: the Fortune 500 companies as well as individual adults. They've managed to attract Nobel laureates to design courses and the universities have formed spin-off for-profit companies specifically to develop online programmes. This facilitates the commercialization of software and other products, and is way to take a commercial approach to continuing and professional studies without compromising the University's standing.

Then there are the freestanding for-profit virtual universities which are arousing the ire of institutions that have prided themselves on a long history of public service. The most quoted exemplar is Phoenix University, the largest private outfit in the U.S. Now owned by the Apollo Group, it operates the country's largest online programme with 12,200 students. The university tracks student progress and contacts those who don't submit assignments on time or fail to enrol in subsequent courses. Many critics question Phoenix's blatant commercialization, but few doubt the university's impact on continuing professional development provision.

Although e-learning is in its infancy its impact can already by ganged. Now providers are coming on the market all the time and the trend is accelerating to the point of upsetting universities' virtual monopoly in educational accreditation. An information technology training course offered or accredited by Microsoft has undoubtedly become more valuable than a Bachelor of Science from a renowned university.

The more consumerist the approach of the education provider, the more what is taught is influenced by demand. MBAs dominate e-learning provision and IT courses are a close second. While the new consumer/learner demands flexibility, choice and just-in-time learning opportunities, suppliers will inevitably arise who are focused on meeting the demand at the expense of quality and value. And the consumer really the best judge of what course material to choose? Education is a more complex 'product' than toothpaste or washing powder. A totally consumer driven education market is unlikely to be in society's best interest in the long term. The commercialization of education usually goes had-in-hand with desegregation: course design, delivery, tutoring assessment and accreditation may be carried out by different organizations. Students might study courses or modules from different universities or providers and then put themselves forward for examination and accreditation by yet another institution. While most academics loathe marking assignments, they regard this scenario with horror, and blame commercialization for the demise of the community of scholars' concept of a university. The death of the 'course' has also been predicted, with learners—especially corporate and one-the job learners—demanding short study modules. What then happens to the ability to get an overview of a field when learning consists of the students selecting a whole series of unconnected learning 'bites'? Learners will be 'zapping' between short sequences or presentations much as they do between television channels.

But while some faculty view e-learning with alarm, technology-based learning is where most of the pedagogical innovation is taking place in universities. Multimedia learning resources and interactive simulations are being developed for the web. Collaborative learning activities, new forms of online assessment and small group teaching technologies are making online courses more stimulating, interactive and attractive than many face-to-face taught courses.

Despite 'doom and gloom scenarios', most moderate observers of the scene see a continued future for the campus university, especially at the undergraduate level, while e-learning will above all cater to adult professional and independent learners. Some commercialization of education is good if it fosters innovation, concern for quality and responsiveness to consumer demands. But if some is good, more is not necessarily better ! Not in education at least.

5

Fleeing the Dot.Com Era

The internet has been proclaimed as the supreme network, the place where one day all human beings will communicate. So who are these strange people switching off in droves? Two years ago, one of the scientists behind the original computer language that gave birth to the internet announced a daring new plan. Work would begin to spread the network beyond the boundaries of earth: the Inter Plan Net, as it has been baptized, would bring online computing to outer space.

Outlandish as it may at first seem, the plan is consistent with the hype and breezy forecasting all too often surrounding discussion of the internet. Even through mounting evidence points to a stubborn "digital divide" separating not just poor and rich countries but also groups within nations, policymakers, business and computer scientists downplay the figures as brief blips in the unstoppable onward advance of the network.

A figure of a hundred computers per human is not entirely unreasonable, leading to a thousand billion computers in the internet in 2020. This technological optimism has come under increasing scrutiny over the past year. The "Digital Divide", a UN Development Programme report revealed that less than 15 per cent of the world's population accounted for more than 88 per cent of the internet's users. The financial failure of multiple dot.com firms and a sense that the technology is still too slow and unreliable have dented some grander hopes for an "information superhighway." Yet it is a little known and barley noticed trend that may perhaps prove the most troubling; for the first time evidence has emerged of a widespread tendency to abandon the internet.

No forecasting has yet taken account of this. A cast-iron premise in all extrapolations of the internet's future is that once the network is available, prolonged use necessarily follows.

Recent empirical work has suggested otherwise. Cyber dialogue, an internet research consultancy based in the U.S., has uncovered evidence of a slowdown in internet growth based in interview with 1,000 users and 1,000 non-users. They argue that the rate of growth is decelerating overall, and that an absolute decline in the number of users aged 18 to 29 is underway. An interesting claim is that approximately one third of U.S adults simply do not believe they need the internet and what it offers.

The First Sings of a Rebellion Against Commercialization?

Similarly, a 2000 survey in the U.K. found that 40 per cent of the adult population had no intention of going online, while most of them put their reluctance down either to cost or to the belief that the internet was irrelevant of their lives. Only a third of those who had stopped using the internet expected to reconnect in the future. Cyber dialogue estimated there were 9.4 million former users in the U.S., a figure that had jumped to 27.7 million.

People who stop using the internet are poorer and less well-educated. Those who are introduced to it via family and friends are more likely to drop out than those who are self-taught or receive formal training. Most surprisingly, teenagers are more likely to give up than people over 20.

These results must be treated with considerable caution, since former users can easily become active ones at a later date. But the sheer existence of internet malcontents—and in such numbers is cyber dialogue is to believed—raised critical challenges to the entire online industry. Why the decision to abandon use just at the time of giant internet expansion prompted by the introduction of the worldwide web.

Possibly the most salient feature in the trend is how it has coincided with a revolution in the network's character. Though there is as yet no concrete evidence of a link between the two, the suspicion is that the internet's first users have been alienated by the rampant commercialization of the network over the past five years. Perhaps these figures point to the signs of rebellion against the transformation of a computer-linked community into a marketplace.

During its first 20 years, the network was the preserve of computer science professionals, students and academics resourceful enough to negotiate its complicated protocols. But in 1991, the U.S. national science foundation decided to allow commercial traffic, heralding the internet's switch from academic forum to virtual bazaar. The foundation's decision was followed by the creation of the commercial internet exchange, designed to regulate the exchange of traffic between newly emerging and profit-oriented internet service providers (ISPs)

From four computers connected in 1969, the number grew to 188 by 1979, 159,000 by 1989, and over 56 million worldwide by mid-1999. But the decisions of the early 1990s and the emergence of the web-allowing user-friendly multimedia features and full integration of the internet's previously dispersed elements (e-mail, file transfer and information access) also marked a change in style. The temptations of advertising on and profiting from this unique interface soon proved overwhelming. The proportion of internet computer addresses ending with.com or .net, and thus in the private sector, has risen steeply as a result. By 1999, these addresses formed 79 per cent of the internet, up from 47 per cent four years earlier. Over the same period, the public sector share, represented by addresses ending with .edu,.mil,.gov and.int, fell from 48 to 17 per cent. Non-profit organizations also dropped in their share, from five to two per cent. From their very low base at the end of the 1980s, business shot to a position of dominance within a decade.

There is no doubting that the internet remains an incredibly diverse collection of resources. Much of the original public sector ethos has also remained; most sites are free, and a slew of data bases of archives held by companies (particularly the press) are open to whoever wishes to access them.

But it does seem highly probable, if as yet empirically unproven, that the internet's mutation has driven away its first zealots, Back in the 1980s, there was relatively little distinction between the producers of what was on offer on the network and the consumers. If you had the skills and resources needed to access the internet, it was more then likely that you would be able to produce content and even services for it. The ethos was "give some, take some"

A String of Mergers in the Scramble for Supremacy

In the first few years of the web, this interactive culture continued, allowing anyone with a small amount of technical knowledge to become an "internet publisher." The change since then has been profound. Increased emphasis on scripting languages, multimedia and links from the web to databases held by organizations has handed power to skilled, professional programmers now, the medium that was initially conceived and an opportunity for smaller organizations and individuals to create and prosper on a "level playing field" has bifurcated into an industry of producers and a mass of consumers.

This trend was reinforced by the spread of search engines such as yahoo ! and Alta Vista, both of which aimed to help the user navigate through the surfeit of internet resources, Commercial extension of this concept has given rise to the "portal"—one of most important elements in the current web industry. While offering traditional search features, these portals also try to maximize revenue by "click-through" advertising and deals with electronic commerce firms. The result of the scramble for portal supremacy has a string of mergers and takeovers aimed at securing top spot in internet rankings, with leading contenders like AOL, yahoo! and

Microsoft acquiring other firms, boosting content, and linking up to media companies such as Disney and ABC network. Add the era of digital television dawns, this convergence of access and content is certain to accelerate.

There are definitely many users who would not object. Even in its mid-1990s incarnation, use of the internet was plagued by problems of connection, slow movement of data, excesses of information and the off loading of "junk mail." It seems likely that some of those who abandoned the internet did so out of disappointment with the sluggish and haphazard service they received once they had signed up for the promised information revolution.

Such former users could well be tempted back by the latest internet developments. A stress on "pushing" predefined content through portals, sophisticated encryption schemes, secure payment systems and faster connections all promise to make the network easier to use and navigate, while simultaneously reducing variety. Add to this the fragmentation of the internet through its introduction in mobile phone, interactive television and palmtops, and it is clear that the marriage between access and content—between the device that leads the user into the network and what the user sees—may well become much tighter. The result could be absolute commercialization.

The Symbolic Value of Joining the Information Highway

This goes to show that the requirements of different groups of users, and particularly groups of former users, are not necessarily the same. The more commercial the network becomes, the more it strays from its free and interactive origins, the more likely certain users will be turned off. A different group of users, on the other hand, might switch off were it to remain chaotic and technical.

Certain trends may help to bridge this gap. The number and breadth of internet resources is still growing, as is the quantity of local content made outside the United States. Simple text message services sent via low-speed telecoms—the so-called "information dirt track" – could also speed the take-up of the internet throughout the developing world, evading the costly trap of commercial, high-speed service providers.

But a whole set of premises about the exponential increase in use of the internet and the loyalty of internet users are under threat. At present, the symbolic value of having internet access is mainly perceived as a sigh of inclusion in an unspecified, high-technology future. Its rejection can be seen to imply a refusal of the intimidating, exhausting pace of technological and social innovation. The internet may travel to space, but it still has convincing to do on earth.

6

Labour Pains

The Birth of Movement

Cyber-rights and business groups are fighting a proposed cyber-crime treaty. While there is strength in numbers, the groups' diversity may prove too much for the coalition to bear. An electic coalition of civil rights and corporate groups, however, has launched an offensive against the proposed cyber-crime treaty as well as a battery of controversial national laws and international standards.

Not surprisingly, the first privacy campaign emerged in the U.S, an internet stronghold, against the infamous Clipper Chip. According to the government, this cryptographic device would have offered a standard for securing private voice communication. Two government agencies would have held the "keys" to be handed out only with "legal authorization". Privacy-minded citizens quickly saw the dangers of this in light of the federal government's history of illegal domestic surveillance. In 1994, 50,000 people—a hefty chunk of the cybernaut population—signed the largest internet petition of its time against the proposal, which died soon after.

Perhaps the most striking feature in the battle to protect privacy has been the diversity of groups involved. The year 1996 saw a motley crew of immigration groups, gun-owners, liberals and conservatives band together to oppose legislation that would notably have extended wiretapping and allowed for more investigations of political groups. Irish-American and Arab-American associations joined out of concern that they would be more aggressively targeted in the "war on terrorism". Meanwhile, the fear of a more invasive government rallied gun-owners alongside groups from across the ideological spectrum. Once again, by tapping the Internet's power to organize and disseminate information, they shelved the legislation.

But this ad hoc co-operation is based on shaky ground. These was considerable coordination, for example, between civil liberties groups and the industry to oppose the 1994 Communications Assistance for Law Enforcement, Act, which requires telecommunications carriers to modify their equipment, facilities and services in order to comply with authorized electronic surveillance. Yet once the industry received a promise of government funding to implement the law, it quickly abandoned the coalition. Corporate representatives then jumped sides again and sued the government over implementation rules in a controversy that is still brewing.

Old Enemies Become New Allies

With the internet's growth, the privacy battle is becoming increasingly international. Most Western European countries have at least one cyber-rights group a trend that is spreading across the continent and Asia, particularly in Japan. At the same time, existing human rights groups have also started to focus on the internet. All it takes is for a single national government to ban free speech on the web and the issue instantly takes on a global character.

For decades, international bodies like the Organisation for Economic Co-operation and Development (OECD), the Council of Europe and the European Union (EU) have been developing international standards relating to privacy, free speech and other civil liberties issues. Their work has included brokering common rules on data protection and encryption policy to promote e-commerce. While some government representatives have put a stronger emphasis on protecting human rights, economic interests have clearly dominated the debate, strongly influenced by the International Chamber of Commerce, a powerful lobby of industry groups. Today, by pressuring governments bilaterally and multilaterally, the U.S. is leading the efforts to expand surveillance worldwide. This pressure amounts to what privacy advocates call "policy laundering" by pushing other governments into accepting controversial plans like the Clipper Chip, international standards will be developed which will in turn force the U.S Congress to accept proposals it had originally rejected.

To respond with more music to these trends, a new opposition front emerged in 1996: the Global Internet Liberty Campaign (GILC), started by the Electronic Privacy Information Center, Human Rights Watch and the American Civil Liberties Association. The group now represents over 50 NGOs from some 30 countries. GLIC operates by consensus. Member organizations propose specific actions such as drafting letters to world leaders, releasing reports and holding conferences. Member groups then agree to join in the action.

GLIC and groups like the Trans-Atlantic Consumer Dialogue (TACD) are making inroads into the policy processes. Perhaps the most tangible signs of their success are the frequent invitations to participate in OECD meetings. But the movement has just one foot in the door: the next step lies in strengthening the role of NGOs outside the U.S. The problem lies in the old Achilles heel of international movements: a lack of funding.

7

Shh...they're Listening

The journalist who first uncovered Echelon, a major electronic spy network, reveals how international surveillance touches us all. Constellations of giant golf balls can be spotted in the most remote locations across the world, from China's Pamir mountains to the swampy north coast of Australia and atop tiny coral atolls in the Indian Ocean. Between 30 and 50 metres wide, these smooth, symmetrical white domes also loom among rice fields in northern Japan and the vineyards and mountains of New Zealand's Island.

The clusters are the most visible sings of concealed electronic network that watch the world. Each dome is filled with satellite tracking dishes that silently soak up and examine millions of taxes, e-mail messages, phone calls and computer data that keep business and political affairs afloat. Unknown to the communicators, their messages are running through the domes, into computer networks and onto listeners who may be on the other side of the planet.

As the world has globalized and international communications have become central to human affairs, these listening networks have grown exponentially. They are part of systems called signals intelligence or "sigint", operated by a handful of advanced countries.

For many years, sigint networks were secret: discussion of their existence was strongly discouraged or even forbidden

by law in the countries concerned. Now the European Parliament is seriously investigating sigint organizations and their impact on human rights and international trade. Europe is focusing on "Echelon", a system that relies on listening stations in about ten countries to intercept and process international satellite communications. Echelon is just one part of an immense network run by the U.S and its English-speaking allies—Britian, Canada, Australia and New Zealand – known as UKUSA after a secret agreement that created the alliance in 1948. Little escapes the UKUSA network, which intercepts messages from the internet, undersea cables and radio transmissions as well as from monitoring equipment installed in embassies. It even operates in space with a fleet of orbiting satellites.

The history of systems like Echelon is as old as radio itself. The first international scandal over secret listening occurred in the 1920s, when the U.S. Senate discovered that British agents were copying every international telegram sent by American telegraph companies. Today's international networks were founded in the early years of the Cold War, when many western countries began jointly monitoring the former Soviet Union.

Fear not the Word "Bomb"

Who is listened to, and why? Officially governments only admit that surveillance is aimed at commonly agreed perils such as arms proliferation, terrorism, drug trafficking and organized crime. But this is the tip of the iceberg. The main aim is to spy on other governments' diplomatic messages and military plans while collecting information about trade. In fact, in 1992, the U.S.re-adjusted its national intelligence priorities, specifying that 40 per cent would be economic or "economic in nature".

While UKUSA is the world's largest network, France, Germany and the Russian Federation have similar systems. On a smaller scale, so do countries in Scandinavia and the Middle East, including Israel, Saudi Arabia and the Gulf states. The budges of all government sigint agencies probably add up to an annual expenditure of $20 billion, according to my calculations for a European Parliament report published last year.

Despite the extraordinary scale of Echelon and its sister systems, the press has mistakenly reported that the network could intercept "all e-mail, telephone, and tax communications". Nor can it recognize the content of every telephone call. And it is pure fiction that by typing key words like "bomb" in an e-mail, you can trigger as tape recorder in some secret base. For every million messages or phone calls intercepted, less than 10 will be used for intelligence purposes. Most personal communications are ignored except those of "important" individuals, like politicians, top business executives and their families.

The UKUSA network, however, does have the power to access and process most of the word's satellite communications and relay contents to client states. The system provides participating countries an enormous, unfair political advantage, since most developing nations cannot afford the expertise and equipment necessary to protect the privacy and security of their works.

Spying on the Government for the People

News about these systems began to leak out in the 1970s as U.S. intelligence agencies came under scrutiny in the "Watergate" affair, when former President Richard Nixon used electronic bugging against his election opponents. Since then, an increasing number of whistleblowers have revealed the scale and effects of sign spying.

Over the next 20 years, official secrecy relaxed as U.S. Congressional investigations notably turned the spotlight on sigint agencies. In Britain in the 1980s, a controversial ban on trade union membership at the Government Communications Headquarters (GCHQ) boomeranged by shifting attention on its spying activities.

The growth of the public information culture on the internet has taken these developments a step further. Now, even GCHQ and NSA have websites to reassure UKUSA citizens that they are not targets. No such safeguards apply to the rest of the world: these citizens are by default denied the right to privacy. The countries intercepting their communications are free to use the intelligence for whatever they wish.

Such conduct violates the Universal Declaration of Human Rights, and the International Telecommunications Convention, which assures the privacy of international communications. Indeed sigint agencies trample over a long line of treaties.

While individuals probably never know that they have been spied upon, their organizations and countries may pay a high price for it. During trade negotiations, sigint agencies can sweep up the messages of a producer nation to discover their bottom line. Armed with such secret reports, the negotiators for the developed world can force prices down to a minimum. Several governments have recently begun targeting environmental organizations or those protesting unfair world trade.

Even when there are no direct adverse consequences, the mere existence of powerful surveillance systems can exercise a chilling effect on free speech inhibiting political and cultural development.

As these activities have become more controversial, the U.S. has tried to expand its circle of collaborators. Countries like Switzerland and Denmark are currently building new satellite stations to gather and trade spy data with the U.S. But, as the ongoing European Parliament inquiry indicates, public awareness and concern are growing fast. Yet vigilance is not enough: if countries and people are to have equal rights in the global information infrastructure, concerted action must quickly follow.

8

Inclusion or Exclusion

Will the networked economy widen or narrow the gap between developing and industrialized countries? As we move from the industrial to the information age, access to the global information infrastructure for economies, companies and individuals becomes paramount.

The debate about the welfare implications of the information revolution for developing countries has given rise to diametrically opposed views. Some believe that information and communications technologies (ICT) can be mechanisms enabling developing countries to "leapfrog" stages of development. Others see the emerging global information infrastructure as contributing to even wider economic divergence between developing and industrialized countries. The reality is more complex.

A number of trends are broadly recognized as the hallmarks of the information age. First, ICT progress is expected to continue to promote the proliferation of communication networks as the costs of delivering these networks decline and the quality of their services improves.

Second, in a networked environment, the incentives for specialization and outsourcing increase. This puts a premium on flexibility and responsiveness as business cycles shorten and interactions between producers and consumers expand.

Third, electronic commerce is expected to continue to expand rapidly and further contribute to the internationalization of service activities.

Fourth, information flows are at the very core of the globalization process as countries and corporations project power by promoting their own culture and values on a global basis.

These trends suggest that the countries that are better positioned to thrive in the new economy are those that can rely on widespread access to communication networks for their companies and citizens; the existence of educated labour-force and consumers; and the availability of institutions that promote knowledge creation and dissemination.

Income Inequality and Computer Literacy Gaps

The quality and coverage of schooling at all levels are also characterized by significant gaps between industrialized and developing countries. These gaps reinforce income inequality, not only internationally, but also within each nation as the ratio of female to male illiteracy tends to be higher the lower the level of economic development and the benefits of public education are typically skewed toward the richer classes. The gaps are even more dramatic when translated to the field of computers literacy.

Finally, developing countries are ill equipped to implement pro-competitive regulatory regimes. In the same vein, the culture of protection and enforcement of intellectual property rights is often an alien concept. The same applies to reliance on networks to promote transparency and access to government services.

All these indicators seem to point towards a social transformation that will increase rather than diminish economic divergence at the international level. Developing economies would be condemned to fall further behind in the international economic race because of their lack of connectivity and ability to transform the information explosion into a knowledge revolution. Once one analyses the drivers of the information revolution, however, a different picture beings to emerge.

Technological developments are rapidly eroding economic and technical barriers to entry into communication networks. Developing countries can, for example, leapfrog stages of development by investing into fully digitized networks rather than continuing to expand their outdated analog-based infrastructure.

First, by developing a modern information infrastructure countries can reduce isolation and exclusion. Many countries are experiencing fast expansion of cellular telephony as an alternative to inefficient conventional network services. Wireless technology can also provide affordable connectivity to rural areas in a fraction of the time that was required in the past to expand conventional telephone networks.

Second, countries can accelerate educational development by using their information infrastructures for distance education. The costs and effectiveness of such programmes are improving dramatically. ICT is also being used for lifelong learning, opening opportunities for new players in education systems. In developing economies, the dynamism of these new entrants can challenge convential educational systems and play a catalytic role in their transformation.

Leapfrogging Stages of Development

Third, a modern information infrastructure can also be a powerful force for better governance. It can, for example, enhance tax administration, audition and control. Moreover, countries can now automate their institutions administering intellectual property rights, strengthening their efficiency and enforcement capability at a fraction of the costs that prevailed in the past.

In short, the logic of the network economy is one of inclusion rather than one of exclusion. As technological progress continues to push the costs of computing and bandwidth down, opportunities for development-oriented applications of ICT will multiply. Moreover, for those already connected the value of the network increases exponentially as new participants join the community.

Technological Determinism and Government Policies

These considerations point towards a more optimistic scenario for developing countries' participation in the emerging knowledge economy. Although, no doubt, income and wealth inequality may increase in the initial stages of the process, catch-up can also happen at a much faster pace than in the past. ICT spending, for example, grew more quickly in most developing regions than in high-income economies in the 1992-97 period. And countries like South Africa and Brazil already boast a higher share of networked personal computers than most industrialized economies.

These scenarios can be criticized for sharing a common feature; technological determinism. The different outcomes predicted, however, illustrate that they are also influenced by other variables, in particular, government policies.

If, for example, developing countries maintain regulatory barriers to the expansion of network e.g., by favouring monopolistic providers for telecom services—the likelihood of the first scenario increases. In this case, global dualism will be magnified not only across the conventional North-South divide, but also in terms of country-level economic inequality as a small elite of connected people in the South benefit from the global information infrastructure.

One the other hand, if regulatory roadblocks are properly addressed and efforts to promote universal access to value-added networks and computers literacy are implemented, then the possibilities for catching up will proliferate. Participation in multilateral efforts e.g., negotiations conducted under the World Trade Organisation and the World Intellectual Property Organization—can also be used to leverage the process of institutional modernization. Under these circumstances, the benefits of the revolution will be widely disseminated both at national and international levels.

The most likely outcome, however, is a combination of both scenarios in which a subset of developing countries is able to converge with high-income economies more quickly than ever before while others lag further behind. International efforts to promote pilot projects in this field can increase the number of countries in the first category.

9

Net Gains or Net Dreams?

What is knowledge-intensive development? Can the internet support it? Will this support make any difference to the lives of people in developing countries who are disadvantaged or marginalised? How can governments and other stakeholders ensure that their societies benefit from the new network technologies and services? This Chapter probes key implications of knowledge intensive development and the internet. The new information networks can help, but governments and other stakeholders must introduce new policies to dismantle barriers.

Never before has so much chatter been heard in policy and business circles as well as in citizens' and ethical interest groups about the impact of advanced information and communication technologies (ICTs) on the global economy and the social order. For developing countries, these impacts are hard to track because of the rapidity of technical change and the unevenness of network and service availability. Education, training and skills development are failing to keep pace with the spread of the new ICTs, a growing source of anxiety for people in developing countries. Research is yielding contradictory evidence about who will benefit and how. The global information infrastructure is penetrating the developing world and making claims on limited investment capital. Poverty, illiteracy, poor health, under-funded education, and worsening environment conditions are also making big claims on public resources. Everyone hopes that digital ICTs and the arrival of the vast networking potential of the internet will help remedy these ills.

The new ICTs are opening access to a flood of information from local and global sources. Converting this information into solutions to high priority development problems is the big challenge. The potential of ITCs must be harnessed to achieve major social and economic benefits. In principle, new digital networks and services could be used to communicate more quickly and cheaply, bringing the village to the world and the world to the village. The vision is one in which inclusion of people in developing countries in a more knowledge-intensive development process follows from access to services like the Internet.

'Net' visionaries portray a world in which access to the Internet and other information and communication services is all that matters. They acknowledge risks for people in developing countries, but they move on quickly to promote the use of new products and applications which, more often than not, have been designed in ignorance of development realities. The globalisation of markets for technologies and services, the rise of dominant firms like Microsoft, the emergence of a handful of global telecommunication operators, and increasing shortages of skills in key areas, mean that neither the benefits nor the risks of these technologies can simply be assumed. Harnessing networks and services to deliver benefits in developing countries means ensuring that those facilities are responsive to the poorest and most disadvantaged groups and communities. It means experimenting with new partnerships that boost equity, mobilize investment for building capacity and spur the harvesting and sharing of scientific and technical knowledge. It also often means pushing for national or regional participation in the global governance system that steer trade, regulation, and intellectual property and privacy protection.

The new networks and services offer fresh opportunities for global and local change. But too much focus on the technical aspects of electronic commerce and new social applications means that organizational, social and cultural transformations are insufficiently heeded. Decision makers in developing countries need to take measures to ensure that ICTs can be used as 'tools' for social and economic development alike. Users need to accumulate new skills through formal and informal education, and learn how to use new sources of scientific and technical information to tackle problems creatively.

The articles that follow show that knowledge from ICTs can be converted into real social and economic benefits if new approaches like 'knowledge management' are effectively employed by governments and other stakeholders to overcome barriers. The internet is enabling research results to be shared among community workers, businessmen and women, educators, and scientists worldwide. This exchange of knowledge is sparking novel ideas about how to harness ICTs to suit development purposes.

More research is urgently needed on the way ICTs are influencing the activities of women and men, on how skills and capabilities can be built up to tackle local and national problems, and on why some initiatives to use the new technologies and services succeed while others fail. It must draw upon the experiences of people in developing countries. Studies of global market trends and structures for ICT supply are needed, to gauge opportunities for people in developing countries to develop new services for strengthening their economies, creating jobs and reducing poverty.

Although there are substantial risks, potential gains from ICTs are far greater. Governments and other stakeholders should face up to the implications of the 'IT Revolution'. National or regional ICT strategies should be set in place, corresponding to each country's development goals. They acknowledged that the next decades are not likely to see the gap between rich and poor vanish. But they argued that if governments and other stakeholders could find ways to use ICTs creatively, the gap at least could be reduced.

The social and economic exclusion of people in developing countries will not be eliminated by 'technobabble' about the global information society, nor by dreams of 'cybertopia'. However, when people's creative efforts and financial resources are combined to use ICTs to encourage innovative forms of knowledge-based development, there are likely to be substantial benefits. Action must be taken to ensure that visions beget policies so that ICTs bring more gains than losses to people in developing countries.

When Computers Chip Away at Our Memories

Galloping advances in information technology promise to give us instant access to all worlds' knowledge. But how will human memory fare against the rise of the super-machine? If the architects of technology's next great leap forward are to be believed, all knowledge may soon be shrunk to vanishing point. Nanotechnology, or computing carried out at the scale of atoms, is their byword for the future. With its awesome potential, scientists have recently argued, around 11 million 400-page volumes—could be stored and printed for instant viewing on device the size of a human palm.

The ink may still be fresh on these biueprints, but the elixir of portable omniscience no longer seems so far away. Seemingly cast-iron laws of ever increasing computer power, along with the rise of powerful new technologies, appear to point to horizon where all that can be known and remembered can be transferred to machines with which human beings then interact at will. And it is a future that for some is already spelling big trouble for the brain.

Surveys Point to Yawing Gaps in General Knowledge

Computers not only distract us from contemplation of deeper valleys; the discourage from contemplation itself. As surveys repeatedly show, knowledge of history, literature, geography and even current affairs seem to be on a steep decline: 60 per cent of adult Americans cannot recall the name of the President who ordered the dropping of the first atomic bomb, just as 77 per cent of young Britons are perplexed by words Magna Carta. The day of the nano-shrunk library could soon come, but will any of its users be able to remember a single line of poetry?

The connection between these yawing gaps in general knowledge and the information technology is by no means established, but a host of thinkers in different fields are sure the issues is one that will shortly become all too pertinent. At the same time as it helps us and extends our physical capabilities, it diminishes our individual faculties. This is a vital question, one which has been around for a long time.

Good or evil, writing has nevertheless formed one of the main tools in the evolution of human memory. Indeed it is civilization's unrelenting hunger for placing memory in external stores – cave paintings then manuscripts, libraries, printed works and finally computers – that has supported the entire march of the species. Each of these new technologies has helped humans "off-load" their memories Pre-literate societies, for instance, depended on oral tradition for their expertise – a practice undermined by the flaws of overworked brains, though fertile ground for epic poetry. Through the written word, memories were freed from the head: knowledge could be stored for retrieval in books, and then redrafted into the sort of novel and complex codes on which modern society is founded.

Becoming Good Memory Managers

The benefits of storing memory outside the brain are unquestionable, but the invention of printing over 500 years ago followed by the post-war onset of computing have added a new note to the process: that of thundering acceleration. One simple equation has come to embody this. It stipulates that computing power – defined in terms of capacity and speed per unit cost—doubles every two years. The trend has held for the last 40 years. Should it continue as expected to around

2020, a personal computer by that year will have exactly the same processing power as a single human brain. Add the promised marvels of nanotechnology, optical and quantum computing, and machines might reach utterly daunting proportions. One penny's worth of computing circa 2099 will have a billion times greater computing capacity than all humans on Earth.

For many cognitive scientists, relation between mind and machine are already undergoing drastic reconfiguration. "Distributed intelligence" is the new maxim, encapsulating all systems in which individuals and computers mesh to carry out a collective task, whether it be landing an aircraft or tracking share prices. The internet is so far the corwning glory—a system that in principle might combine individual users into a potent group min.

All of this may sound abstract, but the effects on memory are being felt now. Facts and figures no longer take pride of place in school curricula. Within the past tow years, South Korea, Singapore and Hong Kong – havens of rote learning—have debated plans to axe huge swathes of standard classroom study. Experts in education stress that students must learn to be adaptable, skilled in manipulating symbols, able to respond to new situations; in short, ready to deal with the new economy, a realm where the computer is king.

We will need a lot of new skills. We have to become good memory managers. We've moved away from managing a lot in our heads to managing memory devices. We have to devote more space to this executive control and less to rote memory storage.

Nurturing Imaginative Thinking at School

A heavy diet of ready-made computer images and programmed toys appears to stunt imaginative thinking. Teachers report that children in our electronic society are becoming alarmingly deficient in generating their own ideas and images. As Generics observes, human memory is much more than simple information processing. There are, for instance, at least five systems of human memory, making up an inordinately rich web of self-reflexive, interweaving recollection that no computer has even come close to imitating. But if memory is increasingly stored in machines that we then manage for our learning, work and leisure, then how will these systems in the brain fare? And how will imagination, intelligence and understanding—all of which depend on an efficiently functioning memory—be affected? The simple answer is: we still do not know.

Yet one image stalks the debate, it is not the old science fiction fear of malevolent computer but of a citizen without a personal memory to speak of, Bertman, for one, is convinced that boundless electronic information may be the deadliest enemy of human knowledge. It's not just enough to remember where we live, what our birthday is, and the name of our wife – there is more to human personality and identity than just the details we can find inside our wallet.

WTO Negotiations on Basic Telecommunications

Telecommunications is a significant and fast-growing sector of trade in services with a current world market estimated at US $513 billion. The objective of the WTO negotiations on basic telecommunications is to secure market-opening commitments from as many countries as possible, involving the introduction of more competition into this sector by removing restrictions that block or hamper supply of services by foreign companies.

History

In mid-1992, interested participants in the Group of Negotiations on Services (GNS) began to discuss the possibility of extending negotiations on the liberalization of basic telecommunications beyond the completion of the Uruguay Round. It was felt that in view of the rapid advances underway in regulatory regimes and technology and other factors there was little likelihood of achieving a high level of commitments on basic telecommunications before the completion of the Round. As a result of these discussions, the GNS agreed to extend negotiations on basic telecommunications and included the Decision on Negotiations on Basic Telecommunications and Annex on Negotiations on Basic the Telecommunications in the find Act of the Uruguay Round. Also as a result of

Uruguay Round, 48 schedules (which count the European Union as one schedule) contain commitments in the telecommunications sector.

The ministers at Marrakesh in April 1994 adopted the Decision on Negotiations on Basic Telecommunications, which established the Negotiations Group on Basic Telecommunications and called upon it to conclude its negotiations and make a final report. The Decision states that the negotiations "shall be entered into on a voluntary basis with a view to the progressive liberalization of trade" in basic telecommunications and that they "shall be comprehensive in scope, with no basic telecommunications excluded a priori". It contains a standstill commitment, which came into effect on 15April 1994, under which "it is understood that no participant shall apply any measure affecting trade in basic telecommunications in such a manner as would improve its negotiating position and leverage.

The Negotiating Group started work in May 1994. Bilateral negotiating sessions began in February 1995, and every meeting since then has been accompanied by a week of bilateral meetings. The Group agreed in April 1995 that draft offers of commitments should be submitted by 31 July 1995. It was understood that these were initial offers, subject to further negotiation and conditional upon the quality of other offers resulting from negotiations. Some of the participants who have submitted offers have indicated that they hope to be able to improve them as a result of further bilateral discussions and, in some cases, of additional domestic reforms.

Responses to questionnaire on market structure, competition and regulatory issues in national telecommunications régimes have enabled a great deal of information to be exchanged between participants. As many reforms are being undertaking concurrent with the negotiations, a number of participants have also provided the Group with updates and supplementary information. Thus, the Group has kept abreast of domestic developments in the sector.

In September 1995, the Group adopted a timetable for completion of negotiations. It provides for monthly meetings

Each meeting is to be preceded by a work of bilateral negations.

High-Level Meeting

They negotiations received an important political boost on 6 October when telecommunications ministers and senior official gathered at the WTO headquarters to review progress in the negotiations. This high-level meeting of the Negotiating Group coincided with Telecom '95 in Geneva, the largest exposition on telecommunications in the world.

The have noted after the meeting that "there was agreement that in order to ensure the widest possible dissemination of the benefits of open markets in basic telecommunications the process of progressive liberalization should be firmly grounded in multilateral principles". Ministers and senior officials also "recognized that the widest possible participation on the negotiations is desirable in order to secure a balanced outcome in the negotiations, to the mutual benefit of all WTO members, developed and developing alike".

Technical Issues

The Group has discussed a number of technical and conceptual issues. Most of these relate to regulatory measures and how to deal with them in the context of the negotiations. They include licensing, interconnection, competition safeguard, independent regulatory bodies, frequency and numbering, standards and type of approvalm, transparency, tariffs and according rates, termination services, rights of way, and universal service. Participants have considered whether certain of these measures are adequately addressed by general provisions of the GATS and the Annex of Telecommunications or whether they should be addressed in schedules of commitments or through the elaboration of general rules or understandings.

One result of these discussions is a draft model schedule of commitments on basic telecommunications.

12

Users Are Choosers

States rarely manage to force people to use a language against their will. The fate of languages ultimately depends on their speaker's needs just as ecology depicts the world as a series of interlocking parts ranging from the single cell to the ecosphere, the planet's languages can be presented as a system based on gravity. Today the keystone of that system is English, a "hypercentral" language around which a dozen "supercentral" languages gravitate. Between 100 and 200 "central" languages, linked to the supercentral ones through bilingual speakers, are in their turn surrounded by 4,000-5,000 "peripheral" languages.

These languages do not all have the same weight, the same vigor or the same prospects. The future of the vast majority is in doubt, and more and more efforts are being made to preserve them. Languages, like baby seals or whales, are regarded as endangered species.

But anxiety is not only focusing on the "peripheral" languages. Concern is also being expressed about the widely-spoken hyper or super-central languages, including English and French. In the United States, organizations such as US English, US First and Save Our Schools are campaigning for English to be recognized as the country's sole official language in face of growing bilinguals due to substantial Hispanic immigration. In France, the 1994 "Toubon Law" was an attempt to regulate the use of French by resisting the use of words borrowed from other languages.

The Myth of Language Purity

Language purity is a myth which leads only to stagnation. The Latin that Cicero spoke is perhaps a pure language, but nobody speaks it any more. Today, different versions of Latin that have developed over the centuries are used. They include Italian, Spanish, Romanian, French and Catalan.

This myth, this desire to protect, illustrates a thoughtless fear of change, of borrowing words and expression from other languages. It is as if only stability could somehow guarantee identity. How far can or should policies to protect languages go? Is it possible to keep alive language forms abandoned by their users, sustaining them by a kind of drip-feed or other forms of intensive medication?

"Language War": A Convenient Metaphor

A language policy can only work if it is attuned to the way in which a society is evolving. Only rarely can a language or reform be imposed on people against their will. Is it possible to defend (or save) a language whose speakers don't want it any more? The issue is not the language itself but the importance attached to it by its speakers. A language policy cannot ignore them.

A language disappears not only because it is dominated by another, but also—and perhaps above all—because people decide to abandon it and do not pass it on to their children. The term "language war" is a convenient metaphor, but languages themselves cannot wage war on each other. It is people who struggle, fight or agree with each other. We can follow their conflictual relationships by looking at the relationship between their languages.

Linguists are always sorry when a language dies out, but languages are not museum pieces. They belong to the people who speak them and constantly change and adapt to their needs. They are there to serve people and not vice-versa.

The evolution of language forms and the relationships between them is a ongoing process, and while some die, others are born, sometimes before our eyes.

Since the collapse of the Berlin Wall and the break-up of Yugoslavia, new countries have appeared and new languages are making themselves heard-Bosnian, Serbo-Croat. The speakers of these languages are affirming their identity by stressing and increasing the differences between them, though this only amounts to a few dozen words.

These differences, slight though they are so far, perhaps herald a break-up of French, which may become a kind of mother tongue for a new generation of speakers, just as Latin is the mother of the Romance languages. The same goes for English, Arabic and Spanish. The Spanish spoken in Madrid is not quite the same as that spoken in Buenos Aires, and there are differences between London English and Bombay English.

The function of languages influences their form. The languages used for trade in the markets of African capitals are gradually becoming different from their vernacular forms.

In the seventeenth and eighteenth centuries, in various situations, Creole languages developed as a linguistic solution to the communication problem faced by slaves speaking different languages when they were taken to the Indian Ocean islands or the Caribbean. Using European tongues such as English, French or Portuguese as a basis, they created languages which today are different. A Mauritian, a Haitian and a Guyanese cannot understand each other, even though their languages have a common ancestor—French. Perhaps the children of immigrants will one day speak, along with the language of their host country, a Germanicized version of Turkish, say, or a Frenchified Arabic.

An Ever-changing Language Map

English may not escape this process. Its world domination today is indisputable and will probably last for a while. But history shows that the more a language spreads geographically, the more variations it generates, so what happened to Latin may happen to English.

This being the case, the world language map is clearly going to change over the next few centuries. Many languages currently spoken by only a few people are dying out and new ones are appearing or will appear. This means that in the gravitational model described above, languages and their functions will evolve. The hypercentral and supercentral languages may change, and some peripheral languages may become central (and vice versa). Like history itself, the history of languages does not stand still. It moves on, constantly changing and being shaped by the practices of users.

13

The Future of Work

The advent of an 'intangible' economy does not mean the end of work. But it does mean the end of familiar routines and rhythms, of job security, of rigid hierarchies and career planning.

People are worried about the far-reaching transformation of the economy. Are we heading for "the end of work". Yes, we have reached the end of the road. We are no longer creating jobs in industry and automation is sure to reduce their number in the service sector. The quantity of work is thus inexorably bound to decrease.

This thesis may be popular, but it is also mistaken and harmful. History shows that technological innovation has always created jobs on a large scale. In no way is the current trend leading to "the end of work". Just the opposite: the new economy contains huge pools of new jobs which can more than make up for the inevitable loss of traditional jobs.

Dematerialization—the shift away from material products—is revolutionizing all aspects of work—its nature, its organization and its relationship with other activities. Its function is no longer just the manufacture of physical objects but the handling of data, image and symbols. The content of jobs is becoming more abstract. Skilled workers need to know a lot more about mathematics than their fathers or grandfathers did. Even milking cows and manufacturing require more and more calculation, evaluation and control.

Financial Markets that Never Sleep

The organization as well as the product of work is also becoming increasingly intangible. The unity of time, space and action which characterized work in the industrial economy has disintegrated. Work is no longer a regular eight-hours-a-day, five-days-a-week routine. New rhythms have appeared—the hectic pace of financial markets which never sleep, the ups-and-downs of life in show business and the uncertainties of "just-in-time" production where components are delivered a few moments before the final product is assembled.

The new jobs are quitting familiar workplaces such as factories, offices and warehouses. Telework is increasing. Europe's teleworkers may number 10 million by the year 2000, up from one million in 1994.

This upheaval of worktime and workspace is going hand-in-hand with a functional explosion. The range of skills and types of work is expanding all the time. In the United States, the number of job categories has risen from eighty in the 1940s to nearly 800 today. At the same time, traders are dying out faster and faster, especially in information technology where many jobs have a short life of only a few years. Jobs are becoming simultaneously more evanescent and more pervasive, more dissociated and more integrated. On the one hand, fragmentation in time and space seems to be more extensive that it was in the industrial economy. On the other, information technology is strengthening the links between different stages of work and creating an overall fluidity.

Disparities in Productivity

The new forms of work are non-linear. When handling information, knowledge and feelings, there is no direct relationship between the amount of efforts and the final result. This makes for very wide disparities in productivity. In industry, the ratio of the performance of an average worker to that of a good one is no more than one to five. But in immaterial work, an excellent programmer can be a hundred times more productive than an average one.

Non-linear work means non-linear organization. The notion of a rigid, formal hierarchy based on unchanging criteria no longer makes much sense. All that matters now is technical, scientific or artistic skill and the ability to establish a solid relationship with the customer. Functional hierarchy is replaced by "brainpower" – authority gravitates to those who create and control the new stock of intangible assets: data, brand image, technological know-how and human capital.

The new techniques for managing human resources are individualizing the assessment of performance. Two people doing the same job may have different salaries and different status. Automatic across-the-board pay rises are being dropped and replaced by bonuses linked to results. There are no sinecures in the new business enterprise, either for rank-and-file employees, supervisors or technicians—the supposed beneficiaries of the new knowledge economy

Business leaders are no longer a protected species. The head of a big American firm is ten times more likely to be sacked for poor performance now than was the case twenty years ago. The notions of loyalty and of indissoluble links between a firm and its employees are losing their meaning.

The changing nature of work has led to a big increase in so-called non-typical jobs, including part-time, temporary and flexi-time work and short-term contracts. Almost all the jobs created in Europe between 1992 and 1996 were part-time. This trend worries many observers who see it as hidden under employment or disguised unemployment. But they are overly pessimistic. The growth of non-typical jobs is the result of the convergence of several persistent developments.

Where the New Jobs Are

The shrinking number of jobs in traditional sectors of the economy seems to be a general and irreversible trend. In rich countries as a whole, the share of industrial jobs fell from

28 per cent in 1970 to 18 per cent in 1994. Meanwhile, the share of the services sector grew steadily. Four major new sources of jobs can be identified:

Handling information and knowledge: Computer services, research and development, teaching and training account for 40 per cent of knowledge workers. These high-intensity knowledge activities comprised 43 per cent of all new jobs created in the United States between 1990 and 1995, but only 28 per cent of total jobs.

Information technology: Here there is a shortage of personnel, professional groups are sounding the alarm and calling on governments to help. In the European Union countries, the imbalance between supply and demand is such that half a million jobs are waiting to be filled.

The health sector: The growth of high-intensity knowledge services in this field is related to increased life expectancy and the ageing of the population, and the demand for physical and psychological well-being is also steadily increasing. The growth of expenditure on health is persistent and widespread. For the OECD countries as a whole, this spending grew from 3.9 per cent of GDP in 1960 to 7.2 per cent in 1980 and 8.4 per cent in 1992.

The leisure economy: This has triggered the expansion of cultural, sporting and leisure services. It ranges from amusement parks and rock concerts to cultural events such as opera and major art exhibitions. The products of the culture industries have become mass consumer items. Never before have people read so much, listened to so much classical music or visited so many museums. Information technology is also going to add to this vast range of consumer choice. In southern California and New York, the entertainment and multimedia professions are among the main sources of new jobs.

In the labour market, the increase in non-typical jobs is one of the ways in which employers are responding to the pressures of competition and adapting to a global economy which functions seven days a week, twenty-four hours a day. To cope with the new situation, firms are having to figure out how they can use their workers more efficiently and flexibly.

The growth of non-traditional jobs is also due to changing demand. Consumers want to be able to buy a very wide range of goods and services at the drop of a hat, or amuse themselves any time, anywhere. To meet this demand, shops and places of entertainment have to be open late at night and on Sundays. Technology encourages this trends: the virtual economy of the Internet never sleeps.

The widening range of types of work also reelects long-term demographic trends, especially the greater number of women workers and longer life expectancy. Some see non-typical jobs as a necessary evil, while others, especially women with children welcome the change.

The divide between traditional kinds of work and the new jobs is no longer watertight. People are increasingly switching back and forth between the two categories. In the course of a lifetime, a person may change from full-time to part-time work, from an office job to home office and from the security of a big firm to the adventure of entrepreneurship.

Changes in the nature of work are also breaking down the rigid frontiers which marked off the world of work. The traditionally distinct fields of work, education and leisure, are now interwoven and coexist flexibly in a kind of triple helix of social life.

The emerging intangible and relational economy has a huge potential for growth because it is not bound by the constraints of material scarcity. However, the transition to the new economy is an open-ended process. The state has a key part to play in bringing it about. Governments can slow down the rate of change by making it more painful and more costly

Obstacles to Change

Pessimistic scenarios are still plausible, such as that of an economy which generates few new jobs and is polarized between a small elite and the rest of the population who are marginalized and lie in precarious conditions. There is a big risk that this scenario will come to pass because current laws and regulations, as well as widespread pessimistic ideas about work, are powerful obstacles to change. Optimistic scenarios require a wholesale reform of institutional structures and profound changes in behaviour and attitudes. Such far-reaching changes often run into strong opposition from the social and political establishment and come up against the weight of psychological and social tradition. But the gamble of a new approach to work must be made if the transformation to the intangible economy is to succeed.

14

Solving the Unemployment Problem by Looking Beyond the Job

If you had a job, you worked; if you didn't, you didn't. Having a job meant being employed by an organization in a clearly-defined and stable occupational role, with duties, hours, rates of pay and promotion all more or less standardized. But the job–in that meaning of the world–is a social invention, and a fairly recent one.

The job – the kind that you had, or hoped to get – became a central fixture of life. Its importance was great because it served many needs: For managers and efficiency experts, job assignments were the key to assembly-line manufacturing. For union organizers, jobs protected the rights of workers. For political reformers, standardized civil service positions were the essence of good government. Jobs provided an identity to immigrants and recently-urbanized farm workers. They provided a sense of security for individuals and an organizing principle for society.

Jobs functioned in so many ways that it is surprising how many organizations are now opting for other ways to define and manage work. The second job shift is underway. Its emergence can be seen in the increasing use of temporary and part-time workers and contracted-out services, the changing relationships between workers and management, the growing popularity of self-employment and small business.

Indeed, "de-jobbing" is proceeding at such a pace that many economists, management experts and futurists are now talking freely about the end of the job. Bridges predicts that the job as we now it will disappear entirely – replaced by new kinds of flexible work assignments in post-job organizations – and be remembered only as a quaint artefact of the industrial age.

One reason for the change in work is the economic rules of the survival game among organizations that employ workers. To stay successful in today's hitech consumer economy, business have had re-model themselves into what some experts call "agile companies"—ones that are able to respond quickly to conditions in ever-changing, fragmenting, competitive markets.

The "knowledge worker", whose work involves not simply doing something, but also applying theoretical or analytical skills. Such workers are replacing the industrial labourer as the dominant part of the workforce—and their productive activities are likely to be organized and structured much differently from those of their assembly-line predecessors.

De-jobbing as a result of new technology or the emergence of a service economy is a phenomenon that gets a lot of attention these days; but it is not the whole story. At all levels of society, people are improvising livelihoods that do not fit the industrial-era model. Immigrants to the developed countries, often unable to find steady jobs, nevertheless find places in the new landscape by being mobile, flexible, resourceful and imaginative: they moonlight, work part-time, share jobs, start small businesses. Their lives are often extremely difficult, but they are also instructive to those of us who believe you either have a job or you're out of luck.

It is too early to evaluate the implications of this multifaceted transformation of work, or to dismiss it as simply good or bad. Nevertheless, one cannot deny that it is taking place, and will bring about dramatic social changes.

On the downside, the job shift is causing great hardships for many workers and their families. It poses serious challenges to policy-makers, political activists and labour leaders. The basic question appears to be whether the key to global employment—development strategy is to play "catch—up"-trying to bring millions of people around the world into jobs in industries and the public sector; or to play "leapfrog" – creating new forms of employment.

The proposal to generate more employment in agriculture, for example, is based on new demand for agricultural exports from developing countries. The policies designed to make the most of this opportunity include measures to upgrade technology, raise productivity, ensure the supply of essential inputs, establish marketing and distribution channels, create links between agriculture and industry, and cater to export markets.

The issue of part-time work, another kind of employment that is seriously undervalued in the traditional industries – era job mind-set. Part-time work may not offer much at this point to developing countries, where many people are under-employed and wages are low, but it can be of great help in more advanced economies. And it is likely to be a big part of the global work picture in the years ahead.

A certain agility may also be necessary in agriculture, particularly in countries that for many years have depended heavily on producing commodities such as sugar for export as a means of generating income and employment. As Northern laboratories develop non-agricultural substitutes for many of these commodities—and this is already beginning to happen—the bottom may fall out of "monoculture" economies, only economic, but will have long-run political implications as communities attempt to reorganize themselves in response to the changed conditions. It is, therefore, in the interest of raw materials exporters to closely monitor current trends in biotechnology and the use of genetic resources and modify their internal policies in anticipation of potential long-term effects".

This calls for flexibility, and an ability to get information and to act on it. Government officials, development workers, community leaders and individuals will, in some respects, all have to be "knowledge workers" if they are to keep ahead of global changes. Jobs are going to be created no just by putting people to work, but by finding—or creating—new niches where they can be productive.

It is still possible to talk about jobs for all, and to resist the assumption made by many economists that high levels of unemployment are now inevitable. But, as we move ahead into the global information economy, we may be moving back into an older conception of the job, and seeing it again as something you do, rather than as something you have—or that has you.

15

The Price of a Good Read

The price of books in recent years has pitted small independent bookshops against the big chains, supermarkets and the internet. But what is in the customer's best interest? The end-of-year holidays are nearing. Imagine you have decided to give books as presents to all friends and relatives, who are interested in just about everything. The decision is made, but where should you buy them? The standard choice is your local bookshop. There you will probably find a professional bookseller who will be able to recommend the most interesting books for each of your friends, tell the difference between a paperback and a pocket edition, and order from the publisher or distributor any book he or she does not have in stock.

If you live in France, Spain, Germany or any of the other six European Union countries that have a single-price policy for books, it will cost you exactly the same whatever bookshop or department store you go into looking for the cheapest price. And if you are a good customer, the bookseller may give you a five per cent discount, which is usually the maximum allowed under the fixed price laws.

This system is founded on two basic principles: that the same book should cost the same wherever it is bought, whether in the centre of Berlin or the only shop in a tiny Bavarian village, and that unlike shoes and clothes, the price will stay the same all year round because seasonal discounts are not allowed. Under this arrangement, the publisher usually sets the price of the book, giving the bookseller a profit margin of about 30 per cent.

But if you live in Belgium, Sweden or the United Kingdom (where book prices were deregulated in 1995), you will find big differences from shop to shop because the booksellers themselves can set the prices. Although this seems at first sight to benefit the reading customer, the book trade is divided over it. In a world of disappearing national borders, transnational authorities such as the European Union are therefore seeking a degree of standardization which, while obeying the laws of competition, is fair to everyone.

The advocates of the fixed-price system hold up Germany as a example. With its 7,000 or so bookshops and more than 1,200 publishers, it is one of the driving forces in the print world. Of France where, unlike the film industry's dependence on government aid, publishing is self-sufficient. In both countries, all those involved in the book trade—authors, publishers, distributors and booksellers—ardently defend the fixed-price system. Similar attitudes can be found in Spain, where the government has deregulated discounts on textbooks for the current school year, unleashing noisy protests from writers, publishers and booksellers.

Under a free-printing system, the law of supply and demand will immediately affect prices, since publishers mark up their best-sellers to compensate for losses from the discounts handed out by supermarkets. The temptation becomes strong just to publish things that will sell easily and quickly, with a resulting loss of diversity. Furthermore, for every 200 books published there is rarely more than one best-seller, and it is not unusual for these to keep on turning up in the catalogues of the same handful of publishers—namely those who can afford to pay huge advances to the most famous authors and buy the most expensive foreign rights.

Of course books are not the only products with fixed prices. In virtually every country, cigarettes, medicine and newspapers are as well. Books are a cultural product and as such they deserve every protection we can give them, especially as they are now threatened by things such as pirating and illegal copying.

But let's continue our book-buying expedition. If you have a credit card and an internet connection, you can order your books from one of the many online bookshops, which will send the books to your home. You will probably have to pay the shipping charges and trust the efficiency of the postal system, but the advantages are clear: you can buy the books without leaving your home or office from businesses that are open round the week.

Lastly, the other factor that can make your holiday purchases cheaper or dearer is tax. If you live in Sweden or Denmark, book prices may not be very expensive, but you will have to add a 25 per cent sales tax, something that does not exist in the United Kingdom, Ireland and Norway.

So what is the best way to buy your books? There is no single answer. Do you what to give best-sellers, books about current affairs, reference books, original works or translations? Do you want prose or poetry? Full or abridged versions? Complete works or anthologies? Hardback or paperback?

Whatever price you pay, dear reader, the future of books is in your hands. Enjoy them.

16

Will Education Go to Market?

The World Trade Organisation has launched processes that could open up to competition the expanding and highly protected world market in education. What issues are at stake? Most of us see education as first and foremost a public service which is responsible for providing young people with instruction. For investors looking for somewhere to put their money it is also annual budget of $1,000 billion worldwide, a sector employing 50 million people, and above all a billion potential customers in the form of students.

The decision to extend to services the liberalization of international trade which previously applied to commodities was taken in 1994. The General Agreement on Trade in Services (GATS) which was signed in April of that year included education on the list of services to be liberalized. To say outside the scope of this agreement a country's education system must be completely financed and administered by the state, which is no longer the case anywhere. However, each country can still decide freely what commitments it wants to make, and especially which educational sectors it wants to expose to market forces. The New Zealand government, for example, has decided to open up to outside competition the whole private education sector, from primary to university level.

So far, New Zealand is an exception, but that situation is likely to change. Part 4 of the GATS agreement

("Progressive liberalization") requires that fresh negotiations should be held by the end of 2000 at the latest, and should be directed to "the elimination of the adverse effects on trade in services of measures as a means of providing effective market access". At the Geneva headquarters of the World Trade Organisation (WTO), far from the headlines and the demonstrators, work still goes on. But independently of the WTO and national policies, a number of factors are driving educational systems towards "communication".

Pressures for Change

First, education is a rapidly-growing sector in which governments are finding it harder and harder to satisfy demand, above all in higher education. Between 1985 and 1992, the number of students in higher education rose about 26 per cent—from 58.6 to 73.7 million. Meanwhile, public spending on education has tended to stagnate over the past 15 years (5-6 per cent of GDP in rich countries and 4 per cent elsewhere)

In view of this dearth of public spending, parents and students are increasingly looking to private education for a solution. In the United States every episode of violence in a state school and every scandal that rocks official school systems gives a boost to "home schooling"' where children no longer attend school and are taught at home.

Traditional public education is also coming in for strong criticism. Employers complain it is not geared to their needs and is not flexible enough. Under pressure from economic interests, a process of "deregulating" education systems has begun. The growing independence of schools is encouraging them to look for alternative sources of funding, ranging from sponsorship to full management by private companies and including many kinds of partnerships between schools and firms. The time for out-of-school education has come...the liberalization of the educational process thereby made possible will lead to control by education service providers who are more innovative than the traditional structures.

The development and spread of information and communication technologies on a massive scale make possible the development of paid distance learning, using multimedia and the Internet for tutorials, examinations, etc.

Secondary and primary education are also affected. More and more paying internet sites bill themselves as alternatives to state schools or traditional private schools. The computer screen takes over from the teacher for a fee of around $2,250 a year.

The WTO secretariat set up a working group in 1998 to look at prospectus for more liberalized education. Its report pointed to the rapid growth of distance learning and noted the increasing number of partnerships between educational institutions and private firms.

Education for Export

Some 350 U.S. experts on international trade in services, including 170 businessmen and women, gathered at the U.S. Commerce Department in Washington on October 16, 1998 to draw up recommendations for the U.S negotiators at the WTO. The purpose of the meeting, called Services 2000, was to look at how the U.S Government should continue to support the efforts of American business to take competitive advantage in foreign markets". The U.S. currently controls about 16 per cent of the world market in services. Its services exports have more than doubled in the past 10 years and now cover 42 per cent of the non-services trade deficit.

The United States is also the world's leading exporter of educational services, and a working group at the Services 2000 conference paid special attention to this sector. It concluded that the sector "needs the same degree of transparency, transferability and interchangeability, mutual recognition, and freedom from undue regulation or restraints and barriers that the United States acknowledges on behalf of other service industries". The report said that three points should be at the centre of WTO negotiations about education.

Firstly, there should be a free flow of electronic information and means of communication, nationally and internationally. Secondly, the negotiators should tackle "barriers and other restrictions that limit or prevent the provision of educational and training services across countries and internationally". They were also to deal with obstacles to the transferability of degrees and diplomas.

Fighting for Market Share

The U.S demands are backed by most countries of the APEC (Asia-Pacific Economic Cooperation) zone. In a note in October 1999, the Australian delegation to the WTO said it would be "encouraging all members to make expanded commitments in all sectors, even the ones that have proved difficult in both regional and multilateral services negotiations: particularly education.

South Korea took a similar position. At a meeting of ministers of human resources from APEC countries that it hosted in September 1997, the Seoul government put out a memorandum which clearly stated its vision of education as a tool of economic competition.

"The emphasis on education for itself or on education for good members of community without a large emphasis on preparation for future work is no longer appropriate. Such a view of education ad work cannot be justified in a world where economic development is emphasized.

"At present, in many economies, the education systems do not sufficiently reflect labour market conditions. Their inflexible and inefficient education systems could not meet the new economic environmental challenges." So education should be made more "flexible", i.e. be deregulated and liberalized. In particular, "School systems should be established to allow all students to study what they are interested in" and "employers, with school educators, should share the role of educating students".

Some think resistance to liberalizing education will come from Europe, especially France. "The future WTO negotiations cannot call in question France's tradition of public service in the field of education and health," stressed a report on the WTO.

17

Beyond Economics

Unless policymakers take a more all-round view of education, they risk sending their countries down the wrong path. Over the past decade, educational change in most countries has been driven by one imperative: survival in the global economy. This process has been particularly salient in the Asia-Pacific region following the drastic shock of the 1997 economic downturn. But in the current reform process, marked by speeding commercialization and economic preoccupations, other educational missions are being ignored, and countries risk paying a high price for their shortsightedness.

There's no denying that economic considerations are critical in today's world. Students have to acquire the knowledge and skills to survive and compete in the global economy, especially one which more than ever before prizes human capital. A high-quality labour force gives nations a cutting edge in global competition. Understandably, stressing economic returns in the current educational debate attracts private resources. But education has other functions that are the indispensable corollary of more balanced, equitable development. They deserve to be briefly explained.

The first is a social function: education has a role to play in facilitating social mobility and bringing about integration in often very diverse constituencies. It is at school that children learn how to form a broader set of relationships, to live together and become aware of belonging to teach us civic attitudes, to make us aware of our rights and responsibilities-in essence, to become responsible citizens. The task is fundamental in light of democracy's advance in so many countries over the past decade or so. Then there is education's cultural function. Developing creativity and aesthetic awareness, accepting other traditions and belief systems while valuing our own are all part of the path towards fulfilment. Finally, education is a goal in and of itself. Schools help children learn how to learn and play a pivotal role in transferring knowledge from one generation to the next. I believe that all these facets of learning are critical for the long-term prosperity of our societies. In our globalized, interdependent world, these functions take on a more international character. Everywhere, education has a role to play in eliminating racial and gender biases. Promoting global common interests, moments for peace, and greater international understanding.

Rising above Short-term Pressures to Strike a Harmonious Balance

While education is widely recognized as the spine of the learning society, the complexity lies in striking a balance between these various functions. The commercialization of education that we are witnessing the world over inevitably pushes schools, educators , parents and policy-makers to pursue short-term, market-driven outcomes. Lawyers, bankers and businessmen have an increasingly high profile in educational debates. Following Southeast Asia's downturn in 1997, they were influential in changing the academic mindset. In little time, emphasis has shifted from academic achievement to developing communication skills, creativity, adaptability. In and of itself, this is not necessarily regrettable. The problem is that these skills are all perceived to be at the service of a supreme economic value.

Sounder research will be required to analyze and assess where the current trends are leading us. It is increasingly recognized, however, that unless economic growth is accompanied by good governance, a fair sharing of benefits, better social and environmental protection and attention to culture, it will, sooner or later, lead to unrest. It is through education that this broad spectrum of concerns can be nurtured. Policy-makers who have taken stock of this holistic mission unfortunately represent a minority is today's educational debates and reforms. Their foremost challenge is to manage commercialization, to rise above short-term pressures and to take more ethical stance towards education, a long-term strategic view.

18

Crises Prevention

Can Better Development Planning Lessen the Toll of Civil Emergencies and Natural Disasters?

Even a cursory scan of the world's headlines is depressing: armed conflicts are grinding on in Somalia, Afghanistan and in a growing number of other countries. And the effects of natural disasters are becoming more catastrophic each year. International relief aid, in response to such emergencies, has increased substantially. But how large can these sums of money realistically be expected to grow? With no end in sight to the need for relief, the good-will of international donors is quickly giving way to disillusionment.

This leads us to a second question, which is, where does development fit in this grim scenario? For the development community to remain aloof from the issue of disasters and emergencies is not only politically short-sighted, it also ignores totally the causes and the effects of such phenomena.

Natural hazards such as hurricanes and earthquakes may be impossible to prevent. But they only become natural disasters if people are vulnerable. Why is it, for example, that an earthquake in Khilari, Maharashtra that registered 6.9 on the Richter scale killed up to 325,000 people, when an earthquake of almost the exact same magnitude in Los Angeles in 1994 claimed only 57 lives? By reducing poverty we can help increase the coping capacity of vulnerable populations. Therefore helping people lower such vulnerability is as much a development issues as the environment, or women's participation in development. Moreover, the repercussions of natural disasters go far beyond the immediate casualty list that so transfixes the media. Secondary and longer-term effects can be equally if not more devastating. And they must be taken into account by development practitioners.

It has been estimated, for example, that the damage to Mexico City's infrastructure form a massive 1985 earthquake amounted to US$ 3.6 billion. Yet over the subsequent five years, the negative ripple effect on that country's balance of payments resulted in a loss of $8.6 billion. Furthermore, reconstruction requirements forced Mexican authorities to revise their economic policies to meet an increased demand for public funding, credits and imports. The priorities for public expenditure were redirected to reconstruction projects, leaving many of the pre-disaster problems of the city and its people unattended.

In Bangladesh, floods in the recent past claimed 2,000 people. But on closer examination we find that the toll was much more exensive than that: in each of these years the country's economic growth rate was halved by the delayed planting of rice and the destruction of seedbeds in the floods, further undermining the country's food security. All of these are considerations that go beyond relief, but they must be taken into account by development professionals.

Other emergencies may be more complex, but must be subjected to the same analysis. As the situations in Angola, Burundi, Somalia and the former Yugoslavia demonstrate, we know little about the dynamics of emergencies that arise from civil conflict. We do know, however, that their cause usually lies in a lethal mix of poverty, poor governance and ethnic or religious rivalries exacerbated by profound social inequities. We are also learning that their resolution frequently requires the application of peacekeeping and political measures, combined with relief and development.

Among the most virulent effects of such complex emergencies is the massive displacement of people; women and children are the principal victims, constituting 7 per cent of the world's refugees.

These complex emergencies around the world could easily get worse before they get better. This being said, carefully designed development efforts—carried out as building blocks to national reconciliation in the fragile post-conflict stage will need to increase commensurately. The appropriateness and the sustainability of these development efforts will be one of the most important factors in determining whether peace itself becomes sustainable. For example, the absence of carefully tailored reintegration strategies for demobilized soldiers and their host communities would be an almost open invitation to resumed violence.

Yet we must also be conscious of the impact of aid and try harder to prevent the need for relief in the first place. An increasing body of evidence suggests, for example, that emergency aid can sometimes be counter-productive in the longer term, increasing the vulnerability of populations and impeding recovery. Ironically, we find ourselves in situation today where it is far easier to obtain funds for maintaining refugees in their places of asylum than for helping them reintegrate into their own societies. In such cases, we may very well be helping to perpetuate the problem that we sought to relieve, as the presence of large numbers of refugees is sometimes itself a cause of conflict.

So how are we to proceed? And what exactly is the nature of the relief to development continuum that remains logical in the abstract but elusive in reality? The concept of a continuum does not imply a linear and absolutely progressive set of responses. On the contrary, it means that we are dealing with a set of processes rather than rigidly defined steps. It also means that development must be very much part of the disaster management process, and that the aim of the continuum must be to move from relief to rehabilitation and resumed development at the earliest opportunity. However, this resumed development must include conscious measures to reduce the vulnerability that caused the disaster or the emergency in the first place.

In other words, we must give greater thought to prevention before we reach for the "cure" – for humanitarian, political and financial reasons. (The Japanese insurance industry spends $ 200 million a year on disaster education alone). And as development paractitioners, we must reconcile ourselves to the vastly more complicated environment in which we have to operate.

This means, for example, that we will have to begin examining whether the economic policy "medicine" often prescribed will reduce conflict or enhance it. We will have to ask ourselves if the reconstruction period following a civil conflict or natural disaster is the right time to advocate cuts in social spending, as has happened in certain countries in Africa and Latin America. Similarly, is it really in children's best interests to build a school in a seismic zone without first ensuring its structural stability? And does it really make sense to urge drought-prone countries to increase their reliance on cash crops, as has been done in some instances.

A story that never made headlines anywhere involves hundreds of the poorest people in Bangladesh, whose homes remained intact during the floods of 1998, when many others were simply washed away. These people were fortunate enough to have obtained credit through the Grameen Bank for construction materials as well as instruction in the building of flood-resistant homes. The Grameen revolving found had received start-up capital from international financial agencies. Since that time the effort has been expanded, and more than 10,500 flood-resistant homes have been built in the last two years.

This is just one example of the kind of action we need more of—in fairly predictable and recurring circumstances such as the floods in Bangladesh, as well as in the more complex, man-made emergencies to which we must respond.

19

Technological Entrepreneurship

The New Force for Economic Growth

Entrepreneurship has emerged as a major new force for change. The dynamic role of modern small business in economic growth has received fresh recognition worldwide. It is essential to promote entrepreneurship and to mobilize the dynamism of the private sector for accelerated national development. An unbridled private sector may not, however, ensure growth with equity. It is the prime responsibility of governments to create policy frameworks that enable businesses to apply technology for competitive advantage and for the well-being of the public.

The Changing Global Environment

As agents of change and progress, entrepreneurs start by identifying a market opportunity and matching this with social or technical innovations. They then proceed to mobilize the resources necessary to drive their business concept to its commercial realization. The development of a product or service with a high-technology content–never easy anywhere, or at today's rapidly-changing global environment. It calls for restructuring the available technology and business development systems and developing the skills needed by a new breed of "techno-entrepreneurs" to transform innovations into market opportunities at home and aboard. It also requires reorienting the present processes and priorities of technical and economic cooperation among countries.

Amidst the global concerns of environmental preservation, poverty elimination and social development, the practical problems of entrepreneurship are not being properly addressed, even though entrepreneurs will create the bulk of enterprises, jobs and wealth.

A torrent of technology-based goods hits the market every week, ostensibly improving the quality of our lives while simultaneously creating complexity and dislocation. The pace of progress in information technologies, microelectronics, robotics, new materials, biomedical sciences, space science and other advanced technologies quickens, significantly changing the way we live. The growth of markets for these technologies also proceeds apace.

Further, technological change is taking place today against a background of growing intra-national and international disequilibria. While the transformation from state-centered to market-oriented development is opening up enormous opportunities and options, it has also caused severe short-term hardships. In order to survive and prosper in these changing times, India and its enterprises need enlightened government policies, good technical infrastructure and strong cultural roots.

Traditional production factors are giving way to a new paradigm characterized by new patterns of trade, investment and employment, and by informal networking life-long learning and technological entrepreneurship. The manufacturing sector in India continues to be dominated by food products, textiles, chemicals and other traditional industry, mainly in the public sector. However, change is coming, albeit slowly. State enterprises are being corporatized pending privatization, and the share of knowledge-based and information-related activities in the market place is rising perceptibly. Restructuring policies now place emphasis (often purely rhetorical) on the role of the private sector. The legacy of decades of centrally-planned development is generally inimical to private enterprise. In turn, the private sector has been slow to respond to economic liberalization in India and generally failed to generate the new employment necessary to absorb new entrants to the labour force.

The regulatory problems of an onerous tax structure and administration, poor access to finance and raw materials, over-regulation of labour and land use, pervasive bureaucracy and restricted markets have been significant barriers to entrepreneurial growth.

Towards Competitive Performance

The imperative of improved performance has serious implications for India if it is to survive, stay abreast and succeed. It calls for national efforts on systemic efficiency and productivity growth, the move from an investment-driven to an innovation-driven economy and sustained higher-order competitiveness; towards enhanced customer satisfaction at home and penetration of selected markets abroad. Concurrently, governments and business have to address such intractable problems as poverty, corruption and the degradation of the environment.

Creating New Technology-based Ventures

Starting a new business in India is a hazardous task. Problems are compounded when the venture is technology-based:

- Capital requirements are generally larger, while traditional banks are ill-equipped to process the perceived risk. Venture capital generally only becomes an option when the venture has documented the merits of its management, market and innovation.
- Knowledge-based ventures can benefit from linkages to sources of knowledge – e.g. the technical university or research lab. Such mentoring needs to be cultivated.
- Techno-entrepreneurs often have technical skills but usually lack the business management and

marketing skills necessary for success. These need to be supplemented.

- In fields where technology is changing rapidly, it is often advantageous to make technology-acquisition arrangements. Sourcing such innovations, negotiating technology licensing agreements and protecting the intellectual property itself require special skills.
- Knowledge-based innovations are inherently more risky than others. The management of this unique risk requires assessment techniques and vision.
- Technology-based ventures often have social and environmental implications, which need to be managed carefully.
- Penetrating a competitive market requires good market intelligence, a good strategic plan and good luck.

Special Characteristics of "Techno-entrepreneurs"

The popular misconceptions are that techno-entrepreneurs are born, not made; that they take risks with other people's money and fail more often than they succeed. In fact, entrepreneur skills can be identified and developed. The entrepreneur is typically an innovator who formulates new solutions to existing problems, mobilizes resources and stimulates others to participate in his or her team. These aptitudes develop over time, often starting in childhood, as the person faces new challenges and learns from failure.

Entrepreneurial opportunities can be found in every industrializing country, community and family. Principal sources of entrepreneurs for knowledge-based ventures are often the university and government research laboratories, the large industrial and military establishments and professional service firms. Some motivations of the entrepreneur are the need to: be independent; create value; contribute to society; earn recognition; become rich or, quite often, simply not to be unemployed. Value-adding ventures with good growth potential can best be developed in an open market and in a culture which supports risk-taking.

The techno-entrepreneur anywhere has the challenge of moving a concept through the prototype and production phases towards creation of a product which meets market needs at a price consistent with the value created and with the ability of customers to pay.

Equally important, the market itself has to be developed and sustained. It is not enough to be first with a better mousetrap if one does not have the skills to educate and reach potential buyers and to set the market standard.

Hence one has to distinguish between innovators and inventors. The inventor is typically a creative person in a quest for knowledge or for producing new products, without determining in advance whether a real market exists for his or her inventions. On the other hand, the innovator draws on existing knowledge and the talents of others to develop or adapt a product or service at a volume and cost that can capture a significant portion of an identified market. The flexibility and creativity of a small entrepreneurial techno-venture may lead to more incremental and break through innovations than can be generated by large-sized firms in many sectors.

The pace and pattern of India's economic development now depend in large measure on its technical resource base. In this context, the key determinants are the skills to apply technology for enhanced competitiveness, as well as to create techbased ventures. Techno-entrepreneurs have to be supported by appropriate national structures and international linkages if they are to survive and flourish in an intensely competitive world.

20

Employment and Promoting Ecology

How a Service Culture Could Put People Back to Work

We are facing two big and urgent social problems, employment and ecology. Both the unemployment of millions of people and the progressive destruction of the ecosphere are alarming. But they are linked with each other. The 'greening' of industrial products, processes and services could proved many more jobs.

Unemployment has many causes, including:

- Sluggish markets;
- Stagnating or declining purchasing power;
- Growing uncertainty about the future at all levels;
- Lack of will and/or ability to innovate.

But joblessness is by far due mostly to the high efficiency of industrial machinery, which produces ever more, ever faster, with ever fewer workers.

Waste of Resources

The extremely high productive use of human labour and the extremely low productive use of resources are manifested by gigantic mountains of waste. Already today, the junked cars on scrap heaps alone would form a line that would reach to the moon. The scene is the same with discarded electrical and electronic appliances. Every year, millions of tons of ovens, washing machines, refrigerators, dishwashers, TV sets, entertainment electronics equipment and small appliances are being wasted.

If we throw away all these things after a relatively short time we are not only being wasteful and irresponsible with resources, but equally so with people's work. For with the products and materials we discard, we also dispose of the human labour they contain. It is imperative that we radically reduce the enormous turnovers of material and energy. In other words, the productivity of raw materials and energy must be markedly increased. Specifically, that means we must draw as many services as possible from one kilogram of material or 1 kwh of energy. Reducing the enormous flows of materials into the industrial system, as well as developing cycles of materials and responsibility (the manufacturer taken back and repairing and/or remanufacturing used products and materials) are the main pillars of a sustainable development that can cope with the future.

The industrialized nations must cut their consumption of raw materials by a factor of bout 10 by 2050 if they are to be able to handle the challenges of the future. To achieve that reduction, innovation efforts must be directed at increasing resource productivity and/or ecological efficiency. In particular, strategies to extend the useful life of goods and intensify their use could result in reducing both the speed and volume of the flows of resources to industry.

Increasing Resource Productivity

In dealing with nature, we and industry are facing radical change. This is the transition from environmental protection (peservation of nature and health) to greater resource productivity (which at the same time means greater competitiveness). As a rule, environmental protection costs money, while higher resource productivity usually cuts manufacturing costs and/or increases a company's profitability. If the company can sell the same utility or benefits while using fewer resources, it saves twofold: in buying raw materials and on waste disposal. Thereby the rule is that goods and components cycles are more profitable than resources cycles, and that the company which is first in the market gains an additional competitive advantage in terms of a lead in knowledge and image. If a service, or benefits in the form of services, can be sold instead of products, the decoupling of company success and materials flows is even greater.

Impacts on Employment

The two social problem areas of work and ecology have to date been perceived and treated separately in politics, in industry and in our own minds. And, I believe, with little result. The link between the two must be established.

The strategies to boost resource productivity would have considerable impacts on the change in industrial structures, on handling existing product inventories, and on employment. In particular, the strategies would lead to a switch of focal point from a raw materials-intensive and use-value-related service economy. This is where another view of profitability comes in. Business management would no longer focus on value added, but on maintenance of value over longer periods based on the intrinsic value of a product. Expressed as a question, the value factor, which would move to the centre of business thinking and dealing, means: how can the utilization value be improved and sold? How can products be made with as few raw materials and as little energy as possible and create a high benefit as pollutant-free as possible for as long as possible during their entire life-cycle?

With regard to employment, the production of long-life goods would appear at first sight to lead to a reduction in the need for work. In fact, however, the strategies to increase resource productivity have positive net employment impacts. The reason is that saving resources is based in principle on substituting energy by work, rather than the reverse as has been customary to date.

If the useful life of products is extended, that will not only preserve most of the materials and energy they contain as well as the work invested in them. The products will also require a considerable amount of mostly skilled work input. Reconditioning products is as a rule more labour-intensive than manufacturing them. So large-scale reconditioning and repair work increase the number of skilled jobs and at the same time reduces the inflows of materials and energy.

Comparing a car with a life-cycle of 20 years with two others that each has useful lives of 10 years gives a good example. The first car causes an increase in employment per life-year of about 50 per cent in terms of total work input in manufacture, service, repairs and reconditioning while at the same time reducing the energy consumption by half.

Regionalisation of Industry

Extending product service life would also mean replacing energy and/or capital by skilled work, helping to save money to boot. But not only rising costs of disposal, materials and energy would reduce consumption. Increasing transport costs would also mean that carrying all kinds of freight halfway around the world would make less and less business sense. That would result in ever more products and materials being circulated, reconditioned, and recycled or reduced on a regional basis. In turn, that would create regional jobs, and be more profitable as well as more promotive of technology—not only from ecological aspects.

In addition, a way of doing business which encompassed material and responsibility cycles would no longer differentiate between manufacturing and reconditioning, or between marketing and remarketing. The structure of such an economy would be predominantly decentralized and regionalized so that it could adapt itself to the new cycles. It also would benefit from the greater efficiency of the new working practices.

True, jobs would be lost in the sectors of central production, and raw materials extraction and processing. But at the same time, more and higher-skilled jobs would emerge. These would not only be better qualified jobs, but also decentralized because reconditioning, repairs and maintenance must be done near the customer. And that, in turn, would also reduce goods traffic.

In addition, skilled workers would be needed because in many cases of small production runs it makes sense and is also more economical to hire such people. They can work faster and more flexibly—and mostly cheaper—han fully-automated production lines.

There also would be a growing need for maintenance, repairs and reconditioning. More and more people would be wanted for reconditioning, that is, the remanufacturing of old products. As reconditioning involves far more craft work than highly rationalised new production, there would be a positive impact on the labour market if there were more of the former and correspondingly less of the latter.

From Production to Services

Switching to long-life products and changing from selling products to selling use-values would strengthen the current trend of jobs shifting from industrial production to the service sector. For example, if the service of individual transport were to be sold instead of the product car, the company with the competitive advantage would be the one that had a service centre in every town and village, with appropriately staffed workshops and sales or rental facilities.

Enduring change towards a knowledge-intensive and use-value-related service economy would not only mean that more people would be needed to fill jobs. It would offer more opportunities for part-time work, as well as possibilities of employment for older people and the handicapped. People who earlier could not keep up with the pace of working life would be more inclined to return to it. Another impact would be that many companies would reduce their dependence on the world market. They would no longer switch certain tasks abroad, but assign them to their part-time employees, helping them to meet their commitments as self-employed entrepreneurs.

The latter would be accommodated by an ecology-driven fiscal reform which would make massive cuts or changes in subsidies and raise the cost of energy and raw materials consumption. This move would be accompanied by a reduction in income tax and non-wage costs such as social security contributions. The market would thus be more efficient, energy and material-intensive new production more expensive, labour intensive repair work and reconditioning cheaper, and jobs would remain in the home country or region.

A number of more recent studies show clearly that an ecological tax reform would help to create jobs, and thereby could make a decisive contribution to reducing unemployment

Development

The Third Way

While great claims are being made for the increasingly more efficient and effective technologies perfected to serve development of the people in this scientific age, huge problems are threatening the globe. The problems are mass poverty and hunger, underdevelopment, waste, unemployment, resource scarcity, environmental destruction and armed conflict.

In finding answers to the prevailing problems we must be clear about the meaning and purpose of development. The first glaring mistake made is that development is interpreted as development of the economy and not as the total development of society. When economic development is made the supreme goal, most of the other vital aspects get ignored, namely, development of the political system, community, social cohesion, the ecology, culture and values. Development and stability go hand in hand while poverty and chaos constitute the antithetical twin.

Appropriate Development

The key elements in the conception of appropriate development consist of: first, aiming at sufficiently comfortable material living standards and not affluent standards of the rich as in prosperous nations. Second, development must not be confused with GNP growth. Mere increase of economic activity must be not pursed exclusively at the cost of articles that are urgently needed by the poor majority to maintain them at a reasonable level of material living. Third, in the villages, we must produce articles as are needed by the villagers. Fourth, grassroots and participatory development is essential so that the local people identify and solve their local problems. Fifth, instead of capital and energy intensive high technology, use labour-intensive technology, Instead of heavy industrialization, promote medium scale industries and technologies. And, sixth, instead of preoccupation with a high GNP growth rate, focus on the development of communities and of rural bodies and take care to conserve the local ecosystems. The main purpose should be to meet the primary needs of ordinary people and promotion of their productive resources such as land.

In developing countries like India where billions of poor people remain condemned by conventional economic development strategies and theories, it is vital to introduce appropriate development measures to remove deprivation and ensure the necessary of modest living standards.

The world has witnessed the operation of the two systems namely, the capitalist and the socialist one that have obtained in different countries. Although both the systems have underlined the welfare of all as the basic goal, both have left a legacy of waste, hunger and gross human inequality. Some 1000 million people do not get enough to eat including some 20 million in the USA.

Third World Way of Development

The iniquitous situation in the present-day world has sparked off fierce controversies among the conventional economists and the new radical economists who champion a third way, as the alternative way to serve the primary goal of all humanity to have sufficient means to lead a comfortable peaceful life.

In order to achieve prosperity, conventional economists have emphasized the production of bigger cake on the assumptions that everyone will get a slice of it. They also argue that a "tide will lift all the boats". Both these assumptions have proved false in that the poor have neither the slice of cake nor has their boat been lifted. Third way system lays stress on highly localized, less cash-reliant and simply structured set-up. It should not be dependent on transport of goods, but concentrate on more local production to meet local needs with a role for barter and free exchange. He urges a radical re-think of conventional economics.

Is This Stepping Backwards?

The most common criticism levelled against the third way is that it will arrest the progress made hitherto and that it might mean a return to a 'primitive' way of life. There should be no fear on this score because the third way aims at the education in the use of resources and, therefore, of excessive production and consumption. It does not in any sense mean stepping backwards to a lower level of the quality of life. Nor is the alternative way intended to destroy capitalism or socialism. The conception of the alternate way is to promote economic growth compatible with capitalism and socialism. The ground idea is to promote selflessness, mutual concern and social responsibility. This will replace selfish, competitive and avaricious attitudes as have developed in the conventional economic order of today.

NGO'S Resolution at the Rio Conference

That there is increasing awareness of the threat posed by the growing power of multinational corporations was articulated by the international NGO forum in its Declaration resolved on 12 June 1992 at the UN Conference of Environment and Development in Rio de Janeiro. The Declaration states that "the Bretton Woods institutions have served the major instruments by which the destruction policies have been imposed on the world" and calls upon "the world's people to protect their economic, social, cultural and environmental interests against the growing power of transnational capital". The declaration further avers "we recognize the central place of spiritual values and spiritual development...and values of simplicity, love, peace, and reverence for life.

After the failure of the socialist system over four decades to achieve prosperity for all, the Indian Government switched over to the global market economy and is steaming ahead with added liberalization measures to attract foreign investment and the multinational corporations (MNCs). Some adverse effects of this are already visible: for example, majority financial equity granted to MNCs and the emergence of foreign subsidiaries with cent per cent financial equity; the introduction of pizza and Kentucky fried chicken which has been detested by the people in Karnataka. The farmers have also revolted because their rights to produce and sell seeds have been wrested by foreign MNCs who have acquired patent rights over certain Indian crop seeds. In this scenario, the third way has much to commend itself to the Government. The third way has the air of the Gandhian model of economy and production which emphasizes production by people for their own needs and preference for small and medium sized industry. The same paradigm was championed by the renowned economist Schumacher when he said "small is beautiful". India should take good care against the present-day headlong drive for the entry of foreign capital and foreign heavy industries.

22

World Trade–The Next Challenge

On 15 December 1993 the world changed. Maybe not as dramatically as the moment when the Berlin Wall fell, but then unlike that very necessary demolition job, the success of the Uruguay Round was a work of construction. Like the destruction of the wall, though, its effects will be profound and lasing ones felt far beyond its immediate context. It will be seen as a defining moment in modern history.

The importance of the Round can be seen in terms of boost it gives to job creation; to development; to investment; to economic reforms; to the rule of law and in many other ways besides. All of these benefits are real and important. But the true value of the whole is much, much more than the sum of these parts.

Put simply, governments came to the conclusion that the notion of a new world order was not merely attractive but absolutely vital; that the reality of the global market – whatever ambitions some of them may retain for regional in tegration—required a level of multilateral cooperation never before attempted.

No Losers in the Round

It has created a revolutionary framework for economic, legal and political cooperation. But now turn to the immediate results of the Round. Seeing them as a profit and loss account or a scorecard of winners and losers is to see them in static terms, as one-off conclusions with finite effects. This misses the point completely.

Every nation now needs an effective trading system, but especially so the small and poor. They have it. Everyone will also gain from the huge package of market access results even if they did not get every concession they were seeking from trading partners–it is the biggest market access deal ever negotiated.

However, the essence of the Uruguay Round's achievements is that they are dynamic. The new agreements, the new rules and structures it sets up – all mean a commitment to a continuing process of cooperation and reform of which the agreement in December was only the beginning.

Maintaining the liberalizing momentum will call for continuing effort and vigilance by participating countries. But now their energy can be focused through the Round's greatest innovation; the new World Trade Organisation (WTO) in place of the improvised basis on which the GATT has operated for 45 years, trade will now have a permanent forum appropriate to its importance in the world economy.

Technically speaking, the WTO will oversee the implementation of the Round's results, administer all the agreements in goods, services and intellectual property, and manage the unified dispute settlement system. But beyond these administrative functions, it will raise the political profile of trade a profile which has already been lifted greatly by the Uruguay Round. The WTO will have regular instead of occasional–direct ministerial involvement. It will have a clear mandate to act as a forum for further trade negotiations. Most of all it will complete the transition from a trading system which largely restricted itself to policies at the border to one which also covers most aspects of domestic policy-making affecting international competition in goods and services, as well as investment.

Through the WTO, the Round will change the way the world economy is shaped. But it is not the final victory over protectionism and unilateralism. Any premature rejoicing would have quickly been cut short by the evidence since 15 December that major economic powers are still ready to take the unilateral approach to trade problems. Arguments for protectionism based on the alleged threat of low-cost competition to production and jobs will not just fade away because the Round is a success. The seductive appeal of "beggar-thy-neighbour" policies is highlighted by the seemingly greater vigour of the lobbies for protectionism than the advocate of open markets.

These dangers–and the speed with which they have resurfaced–make the achievement of the Uruguay Round all the more important, and its successful implementation all the more urgent. Implementation requires more than mutual backslapping about what we have achieved. It requires now that the US, EU and Japan, in particular, rapidly obtain final authority to ratify and also take a lead in providing the WTO with the means to fulfil its mandate.

The success of the Round has come at a time when it is even more vitally needed than anyone could have guessed when it was launched in 1986. Old structures and alignments have been turned inside out in trade as in every other area of international relations. We face a world of change and challenge, in which the reinforced trading system will be a primary source of stability and security.

The developing countries including India have become enthusiastic supporters of the multilateral trading system and the Uruguay Round even if all their demands were not met by industrial countries. The reasons lie in the changing economic policies of many developing countries and the clearer appreciation of the value of the GATT system that has grown along with these changes.

The challenge of new issues in word trade will be a major one for the WTO. The new organization has to consider issues such as the links between trade and the environment, international competition policy, trade and investment, and trade and labour standards. To say a few words about trade and the environment since it is one area in which GATT member countries have committed themselves already to a comprehensive new work programme. They decided on 15 December, in conjunction with the adoption of the results of the Uruguay Round negotiations, to draw up a work programme on trade and environment by the ministerial meeting in Marrakesh. Environmental policy making is one of the most rapidly evolving areas of national and international policy-making, and it is entirely appropriate that emphasis should be placed now in GATT/WTO on ensuring better policy coordination and multilateral cooperation over the linkages between trade and environment.

Permanent Negotiations

The Uruguay Round may well be the last of its kind but this in no way means the end of multilateal trade negotiations. On the contrary, it means they become a permanent event. Ad hoc negotiating rounds were necessary mainly because the GATT lacked the mandate or the institutional basis to operate the multilateral system to the full on a continuous basis. Between rounds the GATT has tended to lose momentum, often at the very times when it was essential to make the most of the liberalizing impulse. This has allowed protectionism and unilateralism to recover and regroup and meant that each round has to start by regaining lost ground.

The positive results of the Uruguay Round will redefine much more than assumptions about trade. If they are exploited with the same determination, courage and commitment that went into concluding the Round, they should mean nothing less than a new start for sustainable growth and new system of collective economic security for the world.

But if the trading system is now up to the job of supporting multilateral cooperation on such a wide scale, do the other structures of economic cooperation still meet the bill? The establishment of the WTO will put trade and investment on a par–perhaps rather in advance–of cooperation in monetary and financial areas. The WTO will stand alongside its original Bretton Woods sisters, the IMF and the World Bank. The three institutions must learn to work together even more effectively and closely. For example, rather than each body conducting separate reviews of country policies, is there not a case to be made for a more integrated approach on country reviews? But that does not, on its own, add up to effective multilateral economic cooperation. The question really has to be asked seriously: are the G7, the OECD, the regional groupings adequate to provide that cooperation?

It is the next challenge of international economic leadership–the challenge of translating the common interest in global growth into a practical and effective mechanism for solving our common economic problems together. So, the ministers' meeting in marrakesh is an historic event which will establish the World Trade Organization and put in place the new multilateral trading system, they will be making not an end, but a beginning.

23

Migration

The scale and diversity of today's migrations are beyond any previous experience Rapid urban growth and environmental degradation in rural areas have led to internal migration affecting hundreds of millions of people. Migration is now seen as a priority issue equal in political weight to other major global challenges such as the environment, population growth and economic imbalances between regions.

Families and households form the basis for economic growth, social development and personal fulfilment. Decisions, by individual women and men on marriage, family, a place to live, shape the destinies of communities and nations. National policies and international conditions provide the context for individual decision-making. Effective development policies, including population, reproductive health and family planning policies, address this reality.

Data on national and global population trends set the agenda for national policy. An important element of population programmes is gathering data that will allow policy-making responsive to the realities of daily life, and to the needs and aspirations of individuals.

The dominant feature of global demographics is still growth. Age distribution is a growing concern, as the numbers of young and elderly people, grow, relative to the working-age population. The world is growing steadily more urban. From being a sign of strength and dynamism in the national economy, the rate and scale of urban growth has become increasingly a cause for concern. The influx of migrants to the biggest cities may be weakening both urban and rural sectors.

International migration is small in extent compared with internal movements, but has a disproportionate impact. Both internal and international migrations are driven by population growth, and by inequities between countries. Migration is one of the choices which shape people's lives and the destiny of nations. But it can also be a symptom of inequity and underdevelopment. Migrants are by definition the most vulnerable members of the host community. Their living and working conditions should be protected.

Open and frank exchange of information and views between host and sending countries is needed more than ever. The aim of the international community should be to protect the right to move, but to ensure that movement is voluntary and that it stimulates rather than holds back personal and national development. "The point of departure should be the human right to live and work where one pleases, so long as it does not infringe on other people's rights to do the same".

The Urban Transformation

The rural sector is declining in importance and its contribution to national economies. It is increasingly part of a unified economy based on the city. Contact with the urban areas is easier than ever and is encouraged by rural development.

Temporary and circular migration is giving way to more permanent settlement. The largest cities are under increasing strain, and residents are encountering increasing difficulties in improving or even maintaining living conditions. Nevertheless, migration continues, driven by a variety of forces both positive and negative. The choice to move can be part of a strategy for survival or personal development; but it is often enforced by external conditions.

The urban transformation is irreversible, but the rural sectors must also be strengthened to balance the developing economy. Attention to gender issues will be crucial in ensuring a successful transition. The forces driving internal and international migration have much in common. Demographic pressures are contributing to both. As the pressures encouraging migration increase, the options for migrants become more limited. This collision is contributing to the atmosphere of crisis surrounding both urban and international migration.

Costs and Benefits

Migration is the result of individual or family decisions. But it is also part of social process. In economic terms, migration is as much a global phenomenon as trade in commodities or manufactured goods. It is part of a broader pattern, and evidence of changing economic, social and cultural relationships.

But migration may be evidence of a different kind of relationship: the combination of poverty, rapid population growth and environmental damage is a powerful destabilizing factor driving urban growth and eventually international migration. On the recipient side, migration has usually been seen as evidence of a thriving economy: today's industrial states were built in part by migrant labour, skills and investment. In today's increasingly uncertain conditions, migration may be seen as a threat to the security and well-being of the local workforce and society at large.

The only effective means to reduce migration pressures over the long term are to slow population growth; to stimulate economic growth and job creation at home, and promote the development of the individual and the family as the basic economic and social unit.

A Question of Gender

It is often assumed that most migrants are men, in reality, women make up nearly half of the international migrant population. Gender differences in social and economic roles affect migration decision making, household strategy, and the sex composition of labour migration. Attention to the gender dimension of migratory movements ought to be an important component in population and development planning.

Women frequently take the initiative in migration decisions, which may reflect limited opportunities in rural areas. Low status limits women's choices at home and may increase pressure to migrate, but it may also affect life in the host community. Opportunities may be limited by lack of education or skills, or by customer limitation on women's freedom of action outside the family or ethnic group. Paid employment for migrant women is usually in the lowest wage, least secure, and lowest status jobs, mostly in housework, child care and trade.

Most educated women end up in the same low-status. low-wage production and service jobs as unskilled female migrants. Men, too, experience downward mobility, but the contrast in the decline in women's employment status is far greater. Despite these disadvantages women migrants have become significant economic actors. Their status may be improved by migration, but the advantages are not clearcut. Women's status as migrants is affected by their vulnerability, and by their lack of reproductive freedom. To ensure improved status they will need both legal protection and essential services, including reproductive health services.

Refugees

Refugees in the 1990s are overwhelmingly in Asia, Africa and Latin America. Their numbers are large, about 17 million, and growing rapidly. A further 3.5 to 4 million were thought to be in "refugee-like situations", through estimates are probably extremely conservative, and an estimated 23 million people internally displaced.

It is important to recognize the common roots of refugee and other forms of mass movement of populations. At the same time, despite the difficulty of distinguishing between political and socio-economic causes of migration, there is a clear need to distinguish between refugees and other groups of migrants. Participation in international efforts of burden-sharing would ensure that most refugee problems would be dealt with in their regions of origin.

Conclusions and Policies

Migration highlights linkages and interdependencies within countries, with many implications for development agendas, including population programmes and development assistance.

Policies to regulate or moderate international migration have concentrated largely on urban growth. They have been only intermittently effective. The most successful have concentrated on stimulating rural development and the growth of alternative urban centers.

Migration is also a personal or family decision, which is affected by external conditions such as poverty or environmental degradation, improving conditions of personal and family life can make a crucial difference in the decision to migrate, reducing dependence on migration as a strategy. Because migration is the result of personal and family decisions, it can be influenced by policies that improve the quality of life.

This offers the opportunity for policies emphasizing individual development, among them, education, health (including reproductive health) and family planning. Such policies are particularly relevant to the strategies must take into account gender differences in social and economic life and the differential effects of policies.

Migration decisions are about family security and long-term-life-chances rather than simply the maximization of income. They are ultimately strategies designed to look after the individual's and the house-hold's needs, safeguard their security, and respond to their aspirations. If the goal is to reduce migration pressures through development it will be essential to increase the capacity but reduce the need to migrate. Long-term external support will be required to make such policies a reality, particularly in areas of rapid population growth and potential mass outward flows. Highly coordinated allocation of development assistance can help establish priorities and focus attention on basic needs. The challenge to both international donors and co-operating governments is to direct programme spending to the areas where it can be most effective.

Globalization and Knowledge Divide

Globalization looks very different when it is seen, not from the capitals of the West, but from the cities and villages of the South, where most if humanity lives. Four examples taken from India illustrate how the paradoxical forces shaping globalization look when seen from the other side.

Five schoolchildren died in a remote village in India after drinking water and powdered milk mixed in a vat that had contained a powerful insecticide. Nobody could read the label of the vat and children were poisoned. The insecticide in question has been banned in practically every industrialized nation; its sale continues only in places like my country.

Secondly, an important annual event recently took place in North India. Potato growers gather there to exchange the best seeds they have produced in the last year. It is an act of pride for communities to share with others seeds that will help improve the production of potatoes. A transactional corporation attended the festival and are now working to patent the genes of these traditional foodstuffs in order to sell them as profit.

India's macro-economic indicators are excellent. In the offices of investment bankers, you will be told that India is a great investment opportunity. The situation is not so rosy, however. Thirty per cent of the population have been living below the poverty line for the last so many years. Ten per cent of the population are living below the critical poverty line: their income is insufficient to pay for event minimal nourishment. So much of the workforce is unemployed or under-employed.

A distinguished North American political scientist. Dr.Benjamin Barber, recently pointed out that in the United States democracy has degenerated into bringing one group of rascals in for four years, and then throwing them out and replacing them with another group of rascals for four years. From the perspective of the South, that looks very good! In a context where rascals manipulate elections and stay in power for fifteen or sixteen years. I would appreciate the chance to throw them out through peaceful elections every four years.

Thus, the complaints of the North are often the aspirations of the South. Progress in industrialized nations can be a threat to developing countries.

Ten years ago, in the euphoria of globalization and the expansion of services and finance that flowed the fall of the Berlin Wall, I advanced the idea that we were entering a fractured global order. Globalization brings us into contact with one another, but it also strengthens profound divisions and fractures in terms of societies and income, and most importantly in our capacity to generate and utilize knowledge. Over the last ten years, the concentration of wealth and power has greatly increased both within and between societies.

There is a real risk of two civilizations emerging, with two ways of viewing and relating to the world: one based on the capacity to generate and utilize knowledge; the other passively receiving knowledge from abroad and deprived of the ability to modify it.

The world now faces the prospect of this knowledge Divide becoming an unbridgeable abyss. We need the international community to return to the basic principles of international co-operation and introduce the idea that a minimum level of science and technological capability, including access to the internet, is an absolute necessity for developing countries and should be the subject of international solidarity.

This can be achieved. However, contrary to the situation of 20 years ago, national governments are no longer the major players in the game of science and technology. Whether we like it or not, the private sector and the international community of scholars must be invited to the table with governments from the North and South to begin discussing an agenda for the mobilization of a science and technology for development. United Nations with a mandate for the development of the sciences, has a special role to play in the revitalization of international co-operation in this field.

25

Development and Poverty Alleviation

Today the key socio-economic problem is large-scale unemployment. Spreading joblessness brings many other problems in its wake. It erodes national income and living standards, aggravating the already grindingly difficult job of promoting development and alleviating poverty. Joblessness also raises government budget deficits, increasing macro-economic instability while soaking up investment for productive capital expenditure, education, training and relief aid. And joblessness ruins lives and communities by depriving people of the dignity and satisfaction that comes with earning one's keep and making a contribution to the well-being of family and society.

Theories about how best to nurture development (and thus create jobs) have shifted considerably over the last decade. The state role has evolved, in the minds of many, from being a source of relief for the problems of unemployment, poverty and underdevelopment, to being a fundamental cause of these problems through the distorting impact of its intervention on the market.

However, the more market oriented philosophy that grew up during the 1990s has yet to provide convincing solutions in practice at least not on a grand scale, and especially not in terms of job creation as the present jobless economic recovery demonstrates.

The weakness of the current recovery and past approaches to economic development can be traced to the failure to consider employment as the predominant means of promoting growth and alleviating poverty. In policy circles it has too long been an almost ignored priority.

Current trends thus bode poorly, particularly as unemployment rates soar. In light of the circumstances, we need to begin re-examining some of the fundamental questions if only to find out what has gone wrong with the answers.

Minimum Wage?

Let's begin with wages, with corporate restructuring in full force on a global scale, are low wage rates required to raise employment and maximize profits? A top manager of a multinational consumer electronics group certainly thinks so; he likened the perfect factory to a ship "so that we could move it around the world to where labour was cheapest". Perhaps, but this bottom-line emphasis on unit labour costs ignores at least two other factors; namely, that higher wages can act as a screen to select more productive workers and that higher wages translate into better productivity via improved worker nutrition, increased consumption and a generally healthier quality of life.

If higher wages bring these benefits (and it is an open question) should government insist that there be a minimum wage rate? Neo-classical economists tend to respond "no", assuming that a higher wage rate puts money into the pockets of some low wage workers while forcing many others out of work because companies cannot afford to pay them.

Technology Transfer

The impact of technology is another area in need of study. Technological innovation is usually labour-saving and tends to originate in industrialized countries, moving toward developing countries like India, Pakistan where labour tends to be low cost and abundant. Would it therefore make sense to slow down or somehow restrict technology transfer, especially to development markets, in the interest of preserving employment?

The answer here is clearly – no. Historical evidence abundantly demonstrates that attempts to retard technological progress bring about greater poverty and lower growth. Technology, in fact, is at the heart of the new endogenous growth theory which is very much in vogue among development economists today. Slowing down or inhibiting technology transfer would certainly dash many countries' development hopes and aggravate poverty. However, the relationship between technology, development, employment and poverty alleviation is not without its complications.

In the 1980s, the buzz word among development specialists was "appropriate technology" i.e., small-scale and labour-intensive technologies that would increase productive output while allowing an equilibrium solution to be found such that the ratio of the productivity of labour to that of capital is proportional to their relative prices. The conditions for this "small is beautiful" approach to technology tended to be best met in agricultural production. However, where manufacturing industry is concerned, the small-is-beautiful approach foundered badly when the only viable technological alternatives proved to be highly capital-intensive.

Development Gap

A wide gap has emerged between developing countries with an inward focus (which tended to be protectionist and purse policies of import substitution) and those with an outward focus and policy of pursuing export-led growth. Competing in international markets requires technology that is as good as or better than that found in advanced, industrialized nations. Small, therefore, is not beautiful in the global manufacturing economy where product standards are high and the elasticity of substitution between labour and capital is very limited.

The drive to obtain state-of-the-art technology thus leads to a policy conundrum: it is a pre-condition for success in manufactured exports, but the impulse to compete successfully in this most lucrative sector speeds up the transfer of technology from the developed to the developing world, thus reinforcing the bias toward labour saving equipment in developing countries and accelerating a process that is seen as a source of job loss in the industrialized countries.

Technology and Jobs

Before concluding that modern technology transfer is inimical to employment in developing courtiers, we have to distinguish clearly between technology's static and dynamic consequences. In a static sense, it is true that highly capital-intensive export industries may not create much employment on a net basis, but the dynamic effects of technology transfer do contribute to economic growth. And growth, in turn, generates multiplier effects in the form of demand, which stimulates ancillary production activities (like food processing or consumer goods) that rely on more labour intensive technologies.

The problem is that the diffusion and application of technology on a global scale blurs the categories of international product specialization and creates a much more competitive and conflict-prone international environment.

For example, we have already seen the Asian Tigers move from producing goods such as textiles and processed food to producing hi-tech and value-added consumer durables. This advance is only possible due to the growth of human capital (facilitated by investment and higher incomes) and it leaves production of textiles to other industrializing countries, like Indonesia, the Philippines and now China. But the dynamic comes at the expense of jobs in industrialized regions, like the US and the EC, which lost more than a quarter of their work force in textiles during the 1980s. Inspite of job losses, advanced countries continue to produce textiles, notwithstanding major differences in the hourly wage rates for spinning and weaving and the fact that essentially the same hi tech equipment is being used in most production centers.

Protectionism

What has happened in textiles is happening in other industrial sectors (automobiles, for example) as well. The intense market competition is proving to be a source of trade conflicts, and possibly protectionism, as jobs come under increasing pressure.

For many workers and managers, the benefits of foreign direct investment look increasingly like a zero-sum game for employment, and there is a real risk that the tenuous link between overall growth and employment will break down altogether. It is hardly surprising that we are already seeing negatively effected workers and local businesses clamouring for protection in advanced countries.

Government's Role

The concerned governments are supposed to carry out much of this research. The three initial lines of inquiry flow from three reasonable assumptions about the future.

- First, increase in welfare and consumption subsidies are out; investments in training and human capital are in. How can investments in human capital be directed to positive employment effects? Is it perhaps not time to explore more fully benefit schemes targeting the unemployed and the fully benefit schemes targeting the unemployed and the unskilled poor providing them with the type of subsidies that would enhance their human capital, improve their health and productivity through better nutrition and preventive medicine, and restore the dignity of holding a job?
- Second, given the quasi-inevitability of increased automation in manufacturing, how can other sectors (particularly agriculture and services) be developed to export their long-term potential for employment creation?
- Third, given the inevitable pressures of work and productivity in the global economy, what sort of alternative institutional arrangements need to evolve with respect to industrial relations, employment and work conditions?

Finding answers to these and other questions will require no small amount of new thinking, but parochialism or a failure of imagination would be fatal in this global era.

Bibliography

Anand R.P., *Legal Regime of Sea Bed and the Developing Countries,* 1975.

Bhatt, S., *Environment Protection and International Law,* Radiant Publication, New Delhi, 1985, p. 122.

Bhatt, S., *Environmental Laws and Water Resources Management,* Radiant Publication, India, and Advent Books Inc. New York, 1986, p. 355.

Behrman, Danial, *In Partnership with Nature—UNESCO and the Environment* (Paris, 1973).

Bell, Daniel, "Technology, Nature and Society", *American Scholar,* Summer 1973.

Bentley, Glass, "Biology and Human Values", USIS, New Delhi.

Book of Nature: The Way Things Work, Published by George Allen and Unwin Ltd. 1981, p. 525,

Boulding, Kenneth E, "New Goals for Society", S.H. Schun, ed., *Energy, Economic Growth and the Environment.*

Carr, E.H., *What is History,* 1961.

Darlington, C.D., *The Evolution of Man and Society* (London, 1961)

Downing, Paul B., Ed., *Air Pollution and Social Sciences* (New York, 1971).

"Drive to adopt national water policy", *Times of India,* 22 July, 1983.

Dubos, Rene, "Man and His Environment", *Britannica Perspectives,* Vol. 1, 1968.

Einstein, A., *My Views,* ed. by S.K. Bandopadhyaya (Calcutta, 1976).

"Environment Research Programme". Prepared by NCEPC, Department of Science and Technology, New Delhi.

Forbes, R.J., "The Conquest of Nature and its Consequences", *Britannica Perspectives,* Vol. 1, 1968.

Fowler, John M., *Energy and Environment* (New York, 1975).

Fuller, Buckminister R, *Operating Manual for Spaceship Earth* (New York, 1969).

Gandhi, Indira, "Poverty Greatest Pollution, says Mrs. Gandhi". *Times of India,* September, 1981.

Glenn, Seaborg, "Science, Technology and Development: A New World Outlook". USIS, New Delhi.

Hacoley, Amos H., *Human Ecology* (New York, 1950).

"India must develop own ecology". *Times of India,* 8 October, 1981.

Marion, Jerry B., *Energy in Perspective* (London, 1974).

Misra, K.C., *Manual of Plant Ecology,* New Delhi, 1980.

Mukherji, P.K., *Life of Tagore,* trans, by S.K. Ghosh, 1975

Mumford, Lewis, "The Future of Cities", in *Basic Issues in Environment,* E.J. Winn, ed., 1972.

Palmslierna, H., *Future Imperatives for Human Environment,* 1972.

Pavithran A.K., "World Futurology", *Eastern Journal of International Law* (Madras), Vol. 9.

"Plans to usher India into 21st Century", *Times of India*, 24 October, 1985.

Polunin, Nicholas, "The Biosphere Today", *The Environmental Future,* Proceedings of Ist International Conference on Environmental Future in Finland, ed. by N. Polunin, 1972.

Radhakrishnan, S., *Recovery of Faith,* 1967.

Report on the State of Environment, Prepared by Centre for Science and Environment, New Delhi, 1985.

Sarkar, Mahendra Nath, *The Cultural Heritage of India,* Vol. 1.

Sen, Sudhir, "Blueprint for a better World". *Times of India,* 2 March, 1980.

The Limits to Growth, a report to Club of Rome (New York, 1972.)

The Mind of J. Krishnamurti, ed. by L.S.R. Vas (Bombay, 1971).

Toynbee, Arnold, "Man and his Soul", *Hindustan Times,* 4 January, 1968.

United Nations List of National Parks and Protected Areas, 1985.

Vivekananda, Swami, *Complete Works*, Vol. II (Calcutta, 1968).

Ward, Barbara and Dubos Rene, *Only one Earth: The Care and Maintenance of a Small Planet,* Report to UN Conference on human environment, Stockholm, 1972.

"Wildlife Laws in India", *Times of India,* 4 March, 1985.

Ward, Barbara, *Progress for a Small Planet,* 1979.

Index

H

I

J

K

L

M

N

W